ID0850257

The Nature of Being Human

The Nature of Being Human

From Environmentalism to Consciousness

HAROLD FROMM

The Johns Hopkins University Press

Baltimore

The Johns Hopkins University Press
2715 North Charles Street
Baltimore, Maryland 21218-4363
www.press.jhu.edu

Library of Congress Cataloging-in-Publication Data

Fromm, Harold.
The nature of being human : from environmentalism to consciousness / Harold Fromm.
p.   cm.
Includes bibliographical references and index.
ISBN-13: 978-0-8018-9129-8 (hardcover : alk. paper)
ISBN-10: 0-8018-9129-9 (hardcover : alk. paper)
1. Human ecology. 2. Nature—Effect of human beings on. I. Title.
GF41.F76 2008
304.2—dc22   2008024793

A catalog record for this book is available from the British Library.

Page 299 is a continuation of the copyright page.

Special discounts are available for bulk purchases of this book. For more information, please contact Special Sales at 410-516-6936 or specialsales@press.jhu.edu.

The Johns Hopkins University Press uses environmentally friendly book materials, including recycled text paper that is composed of at least 30 percent post-consumer waste, whenever possible. All of our book papers are acid-free, and our jackets and covers are printed on paper with recycled content.

In memory of Gloria Fromm

CONTENTS

ACKNOWLEDGMENTS

From 1968 until her death in 1992, Gloria Fromm was my chief editor and helpmate. Her role in my life and influence in this book are great. My most valuable secondary editor and literary advisor, also departed, was my oldest friend, Bill (William F.) Shuter, who taught at Eastern Michigan University. Beyond the lives and deaths of these personae, the first jolt that generated what was to become *The Nature of Being Human* was the four-year period I inadvisably lived in northwest Indiana, less than twenty miles from the steel mills of Gary (discussed in the early chapters). The second jolt was my meeting with Cheryll Glotfelty in the late 1980s and the force it generated on my understanding of "environment" and "ecology" (dealt with in chapter 5). The third jolt was my introduction to Joseph Carroll and Ellen Dissanayake around 2001 (and the friendships that ensued), which sent me in a new and post-Darwinian direction that incorporated the environmentalism of the earlier periods and inevitably led to my interest in consciousness studies. Their intellectual and editorial presence in my subsequent writings is reflected in the second and third parts of this book, where their own work is specifically examined in chapter 16. And then there's David Barash of the University of Washington, whose expertise has caught a number of imprecisions in some of my rougher drafts. Merely to "thank" all these people who shaped my life seems insufficient.

Other people need to be acknowledged as well. Paula Deitz and the late Frederick Morgan, editors of the *Hudson Review*, have nurtured me for more than twenty years now, teasing out some of my best productions. A few of these are lifted from the journal's pages and duly acknowledged where they appear in the chapters that follow. I must also give admiring thanks to Ralph Cohen, who edited *New Literary History* for so many years and who had the guts to publish two of my wilder writings that more timorous and conventional editors couldn't quite handle. One of these appears here as chapter 20, "Muses, Spooks, Neurons." But if Joe Carroll had not prodded me to attend and give a paper at the conference of the Human Behavior and Evolution Society at the University of Nebraska in 2003, that offbeat (and very seminal, at least for me) manifesto would never have been written and then expanded for publication. Last, but hardly least, Vincent Burke, senior editor at the Johns Hopkins University Press, encouraged and supported me from his first contact with the manuscript, and Gordon Orians, behavioral ecologist from the University of Washington, convinced me (without much resistance) that the book needed a meaty introduction and conclusion, which did wonders in pulling everything together and underscoring the latent themes that run throughout.

Of course, my unwitting neurons deserve the most credit of all, even though neither they nor the fiction naturalized as "I" really know what's going on. Yet they have served "me" well. Without them, nothing.

The Nature of Being Human

From Environmentalism to Consciousness

O N AUGUST 10, 2007, as I read the *Arizona Daily Star* in Tucson, the trigger I'd been awaiting to get started on this introduction offered itself in the headline "Bucolic valley a magnet for urban smog." The Associated Press news item described Arvin, California, a nonindustrial rural area located on the eastern side of the San Joaquin Valley near Bakersfield, as having "the most polluted air in America." Even without its own smokestack industries or heavy traffic, Arvin's geography attracts the flow of pollutants from fairly distant elsewheres, "causing airborne particles to coat homes and streets and blot out views of the nearby Tehachapi range on hot summer days." Moreover, "Doctors and public officials say asthma and other respiratory problems are common among the 15,000 residents who live 20 miles southeast of Bakersfield. People complain of watery eyes, dry throats and inexplicable coughs, particularly in the summer . . . Arvin's level of ozone, the primary component in smog, exceeded the amount considered acceptable by the EPA on an average of 73 days per year between 2004 and 2006." As long ago as 1996, in "Ecology and Ecstasy on Interstate 80" (now chapter 10 in this book), I wrote about the San Joaquin Valley as a cesspool of air pollution. So it's a case of déjà vu. But how did a nonscientist like me pick up such information on my own, in the earliest days of public environmental awareness? Or why should I have cared? As a humanities scholar who has spent his professional life delving into the subtle recesses of tone, style, and esthetic quality, why should the physical quality of the air have been a matter that caught my attention?

A few days later, the Sunday *New York Times* on August 26, 2007, had two long stories about air pollution in China and the forthcoming summer Olympics in Beijing: "As China Roars, Pollution Reaches Deadly Extremes," shouted the main headlines. The World Bank, it was reported, estimates that China is already experiencing as many as four hundred thousand annual deaths from outdoor air pollution, three hundred thousand from indoor pollution, and sixty thousand from toxic water. Olympics in Mexico City, Athens, Los Angeles, and now Beijing? Who are the ostriches who set up these dire venues for athletes, the worst places on the planet for physical exertion? The *Times* article quotes Deena Kaster, an Olympian, who notes that "an athlete is so sensitive to change that even the smallest amount of pollutants has the potential to affect breathing." She adds, "It's strange that we can be such powerful athletes, but be so fragile at the same time." But those 760,000 ordinary citizens expected to die from pollutants will do so without much strenuous effort—why not athletes living at the edge? When Kastor runs in air that is not "clean," she "immediately feels it. Her chest tightens, resulting in an inability to breathe deeply enough to get oxygen to her muscles." The Chinese (in denial of the extent of the situation) expect to curtail pollution during the Olympics by banning cars, etc., etc. An optimist remarks, "It would be embarrassing if they had dozens of athletes having asthma attacks or competing with face masks. China just doesn't want that as their legacy."

The report gives a new dimension to human hubris and the dream world of the mind in relation to the ineluctable materiality of the body. Though the mind can think up daring feats at serious odds with the structure and capacities of the body, the effects of these limitations feed back to produce psychological states ranging from highs of ecstasy-generating exultation to lows of depression, wounded self-esteem, and cheating, drugs, crime, and other pathologies. Yet, the mind habitually denies this subservient relationship to the body, suppresses it, and proclaims itself free. In the chapters that follow, I shall be tracing many particular paths, from the particles and gases in the air through the conditions of the body to states of mind—including my mind, as I think about all this.

If the reports of ecologists make us acutely aware that the physical world presses in on us from without, those of biologists increasingly make us vividly conscious that the material world presses in on us also, from within. A few days after the San Joaquin article, Richard Preston, writing in the *New Yorker* ("An Error in the Code," August 13, 2007), observes that "decades after the discovery of Lesch-Nyhan syndrome, it is still mysterious. It is perhaps the clearest example of a simple change in the human DNA which leads to a striking change in human behavior." This horrifying disease, seemingly caused by one mistake out of three billion letters of code in the human genome, entails wild, self-destructive behavior involving the chewing away of one's own fingers and lips while screaming for help to be restrained. One of the neurologists interviewed by Preston sees Lesch-Nyhan as existing at the extreme terminus of a spectrum of uncontrollable human behavior, whose relatively harmless manifestations may consist of habitually eating ill-advised quarts of ice cream while watching TV, or biting one's nails, or chewing one's lips nervously. But as this moves on to biting cuticles until they bleed, we begin to wonder, at what point of the spectrum does behavior become "compulsive," driven, involuntary? Only at biting off the whole finger and chewing off one's lips? "Where, in this spectrum of behavior," the neurologist asks, "is free will?"

Meanwhile, only another few days later, in the August 28, 2007, "Science Times" section of the Tuesday *New York Times,* John Tierney writes a fascinating report on the universality of the gesture of arms extended with palms upturned. "That simple gesture, the upturned palm, is one of the oldest and most widely understood signals in the world. It's activated by neural circuits inherited from ancient reptiles that abased themselves before larger animals. Chimps and other apes, notably humans, adapted it to ask not just for food, but also for more abstract forms of help, creating a new kind of signal that some researchers believe was the origin of human language." As we extend our arms and turn our palms in a seemingly spontaneous gesture, are we aware of the extent to which this is less an "expression of self" than the automatic enactment of an ancient genetic program that directs us to do so?

It looks as though I could go on in this vein indefinitely. In this new era of evolutionary biology and neuroscience such reports are proliferating at seemingly geometric speed. Where then is the Imperial Self of yesteryear, the unstoppable force of Man's Unconquerable Mind, the miracles of Spirituality, the faith that can move mountains but is unable to stop the metastasis of cancer cells? Even Mother Teresa, we have learned, had serious doubts that God was in his heaven and all was right with the world "he" created, seemingly with the interests of us mortals in mind. Are we just dumb robots? Or as Shakespeare put it, the tennis balls of the gods?

In the chapters that follow, I develop an argument that has appeared in various guises throughout the histories of Western and Eastern philosophy, here applied to the current environmental crises: that the movements to "save the environment" are not as selfless and magnanimous as might at first appear—we're not doing the "environment" a disinterested favor—because the distinction between the environment and us is wholly illusory. If there is only one single matrix out of which everything springs for varying lengths of time and to which it returns for recycling, the urgency of the environment is just one more aspect of Darwinian survival. Species survive or perish as a result of their compatibility with the environment of their time, added to which we as individuals survive or perish as a result of our own personal compatibilities with our immediate life environments, which are aspects of the larger whole. Air pollution, the chemical composition of the foods we eat, the drugs we take: they do not merely produce incidental bad "side effects." They produce *effects* throughout our biochemical composition, because we are derived from the same stuff as they. But the effects are not necessarily benign, not necessarily the ones we have been selected for in the interests of survival. And the environment, already "us," ultimately becomes human consciousness—for good or ill—through Darwinian selection and adaptation, since our very thinking and thoughts are beholden to planetary conditions.

Free agent or robot? Spirit or matter? Mind or body? These hot buttons require fairly firm pressure to be set off in the book's early pages, but by the second half they are triggered by the merest puffs of air. Yet the profound interactions of matter with mind that are implicit in the above examples are hardly novel discoveries. Even someone as "spiritual" as Emerson was moving toward materialism a hundred and fifty years ago (see chapter 18), so maybe what we are now seeing daily in a rush is not new but just a sign of a new urgency produced by new knowledge.

And yet, remarkably, for practitioners of the humanities, almost all of the religions, some of the social sciences, and for much of society in general, the implications from ecology, evolutionary studies, the neurosciences (with their application to both life and literature), and the insights of consciousness studies and philosophy of mind can almost be said to remain terra incognita. Graduate students and young academics in English departments—my incubator as a scholar—tell me that to reveal their interest in these science-derived approaches during a job search is the kiss of death. Literature departments will have little or nothing to do with them, and their flagship professional organization, the Modern Language Association, with the exception of an occasional infiltrating convention paper or session, continues largely to shun them.

It is not as though the humanities disbelieve in evolution or that they resist the general outlook and conclusions of the sciences. They accept the basic scientific doctrines and would endorse the remarkable statement issued by the Department of Biological Sciences of Lehigh University about their own faculty member and colleague, Michael Behe, a proponent of intelligent design. The faculty collectively affirmed that "the validity of any scientific model comes only as a result of rational hypothesis testing, sound experimentation, and findings that can be replicated by others." On the basis of these general principles, the faculty declared that they were "unequivocal in their support of evolutionary theory," and accordingly, they emphatically segregated themselves from Behe. "While we respect Prof. Behe's right to express his views, they are his alone and are in no way endorsed by the department. It is our collective position that intelligent design has no basis in science, has not been tested experimentally, and should not be regarded as scientific."[1]

Few in the humanities would overtly repudiate the idea of evolution, nor overtly support alternatives like intelligent design. Nonetheless, they evade the full force of Darwinian thinking, with its attendant materiality. Treatments of science from within the humanities typically adopt the stance of "social construction"—the assumption that the concepts of science are essentially reflexes of social and ideological forces and do not have any particular claims on "truth" or "reality." Such views on scientific knowledge correspond closely with the general constructivist belief that human behavior is mostly influenced by culture rather than nature. (These are large issues to which I devote more particular attention in several of the chapters in parts 2 and 3.) To discuss Darwinism and modern neuroscience as forms of "discourse" is one thing; actually to use them, to take them seriously, to allow them to shape one's own principles and enter into the inmost recesses of one's thought and language—that is another matter, and it is the matter of this book.

I can anticipate and sympathize with one chief reservation such proclamations are likely to excite among humanists for whom subtlety and nuance are the very essence. For such humanists (myself among them), all Grand Theories seem threatening, monolithic. We naysayers have always looked askance at the tyrannical and totalizing orthodoxy that Freudianism, Marxism, feminism, Lacanianism, structuralism, postcolonialism, and queer studies have exerted over literary scholarship. Would applications of the materiality of the evolved body to the understanding of the arts similarly simplify and brutalize the discourse of the humanities? I can offer only the evidence of my own experience. Materiality has illuminated and revitalized my own sense of the richness of both life and the arts, which I hope this book will exemplify.

Over the past decade or so, my thinking and writing have increasingly been driven by the resistance of the humanities to what they belittle as "scientism." "The New Darwinism in the Humanities," which appeared in journal print in 2003 and is included here as chapter 16, was not intended to suggest that Darwinism had taken over the humanities (far from it) but that it had newly entered their purview. Three years later, in 2006, my review of *The Literary Animal: Evolution and the Nature of Narrative*[2] began by complaining about the ongoing resistance. On the same day I began reading *The Literary Animal,* I had received the 2005 edition of *Profession,* an annual book-length collection of writings about the state of literary studies from the Modern Language Association, the flagship organization for literary academics. It started off with a group of essays about science and the humanities, the first of which was by Louis Menand, a literary scholar and well-known writer for the *New Yorker.* Menand remarked, "Faculty members in science and in social science departments tend to regard humanists as reflexively oppositional to what they do and, therefore, as easy to discount. This perception is founded mainly on ignorance. The summaries of the state of ideas in the humanities in books like E. O. Wilson's *Consilience* and Steven Pinker's *The Blank Slate* are appallingly misinformed." Sounding pretty misinformed himself, he went on make the dubious claim that "the version of the humanities that would make many nonhumanists most comfortable today is the version in which art and literature are ornaments on or neat illustrations of empirical accounts of human life." Though Menand would probably defend the humanities against any charge of anti-scientism, he rejected the idea he supposes to be dangerously prevalent in the sciences, that "human behavior is ultimately understandable in biological terms."

In the same issue of *Profession,* Barbara Herrnstein Smith, a humanities scholar with philosophy and science specialties, nonetheless adopts a similarly hostile and defensive manner, repudiating the "scientism willing to import one or another currently high-profile scientific or sometimes scientoid program into the humanities to make them seem less 'amateurish' and 'impressionistic.'" As I described it in my review, "She alludes with heavy irony to the Sokal Hoax and to E. O. Wilson's (to her) misbegotten program to 'bridge the gap between the two cultures by integrating the anarchic humanities and the floundering social sciences into the more orderly and grown-up natural sciences.'"[3]

At about the time *Profession* 2005 appeared in print, it was widely rumored that "the science wars" were dead. Hardly. They continue even today. In reviewing *The Literary Animal,* I sought to focus the debate on where the real conflicts lie. The contributors came from both the humanities and the social sciences, but they shared a conviction that human dispositions, evolved over long periods of time, powerfully influence the products of the human imagination. Accordingly, I noted, "Far from treating literature as an 'ornament,' the contributors argue for narrative and drama as more or less adaptive. They share a powerful awareness that everything human ultimately derives from the evolved body and brain, no matter how much culture and individual consciousness are capable of varying the forms of expression." It was this awareness that generated the motivating force behind "The Crumbling Mortar of Social Construction," one of the two summings-up that conclude this book.

There are growing enclaves in the humanities—in ecocriticism, biopoetics, science studies, consciousness studies in philosophy, and "literary Darwinism"—in which younger scholars are breaking through the restrictions of the various intellectual establishments. These are like underground rumblings from the margins. But I have been wondering where the first "official" cracks would appear in the wall of resistance within the humanist academy itself. So I was indeed gratified to come across a set of enlightened declarations from within its very heart—the National Humanities Center in North Carolina. Describing a conference scheduled for November 2007, the director of the center, Geoffrey Harpham, overtly acknowledges the massive epistemic transformations that Menand and Smith seem determined still to evade:

A small but growing number of philosophers, literary scholars, and other humanistic thinkers has turned to the work of computational

scientists, primatologists, cognitive scientists, biologists, neuroscientists, and others in their attempts to gain a contemporary understanding of human attributes that have traditionally been described
in abstract, philosophical, or spiritual terms . . . And many feel that
knowledge now being gathered, produced, and developed in various
scientific domains may well force us to modify our understanding of
such traditional moral and philosophical questions as the nature of
human identity; the legitimate scope of agency in determining the circumstances or conditions of one's life; the relation of cognition to embodiment; the role of chance, luck, or fate; the definition of and value
attached to "nature"; and the nature and limits of moral responsibility.

Harpham recognizes that "these suspicions and intuitions have remained
preliminary, and have not yet gathered into a fully formed, or informed,
consensus." Moreover, he is aware that disciplinary differences still present formidable obstacles to any full integration of scientific and humanistic knowledge:

Scientists are often unable to assess the ways in which their work, which
is necessarily limited in its scope, affects human self-understanding,
while humanists are unequipped by training to participate fully in the
generation of new knowledge. The effect on both sets of disciplines is
troubling. The fact that knowledge about the human is being developed in the sciences appears to undercut the traditional function of
the humanities, the analysis, interrogation, and interpretation of the
human. But it also means that the sciences are deprived of a full
awareness of the contexts and implications of their work.[4]

Here we have, in a high-protein nutshell, a summary statement of both
the academic and sociocultural challenges that new knowledge is presenting in the early years of the twenty-first century, often to resisting entrenched interests, not to mention a U.S. Congress and presidency living
in a nostalgic dream world as out of date as acoustically recorded 78 rpm
shellac records for wind-up Victrolas. I don't think there's a crux or challenge mentioned by Harpham that is not taken up in the chapters that
follow. These chapters reveal a dual intellectual evolution—of the knowledge disciplines and of my own understanding, usually as a result of
hands-on experience. What will be seen in their unfolding are the ways in
which evolution, ecology, the "environment," physical matter, the brain,

the self, the mind, and culture gradually merge into one protean substance of variable expressibility.

Although both Harpham and I have focused our attention upon the frequent ignorance and denial of the sciences by the humanities, many scientists (as Harpham acknowledges above) reveal their own blind spots as a result of the increasingly narrow specialties of their disciplines as well as what I have experienced to be a brush-off of the humanities as "soft" and philosophy as simply "metaphysics," even while much of contemporary physics can seem as speculatively metaphysical as Plato and Heidegger. This was made only too clear to me recently at a conference of about fifty science and humanities professionals who virtually lived together for thirteen hours a day over a long weekend, listening to talks ranging from functional magnetic resonance imaging (fMRIs) to music criticism. My most interesting personal discussions were with a world-class imaging specialist and a leading American biological anthropologist. Not only did I draw blanks when I referred to the multiple psychological and bodily effects of air pollution (apart from the generally acknowledged damage from smoking and particulates) but I experienced negative reactions when I brought up the subject of the incoherence of "free will." The science people seemed to think that explanations from quantum physics and stochastics could provide evasions of the materialities of causation almost universally domesticated by post-Cartesian explanations of human behavior. In place of the ancient ghosts finally driven out of the machine, I was offered new spooks, as ineffable and numinous as the old Cartesian ones. But would *chance* motivations from cosmic physics restore willed agency to *Homo sapiens* denied a ghostly "free" self by *deterministic* ones? Since "intrinsic meaning" is an oxymoron, given that all meanings are conferred by thinking beings, even scientists would benefit from the thoughts of meaning specialists, that is, philosophers and other experts of the intelligible. Scientific discoveries and stances, after all, do not speak for themselves but are always already enunciated *interpretations,* which are the playing field of philosophy, linguistics, psychology, and other disciplines of the human sciences. I hope that readers from the physical sciences will find the chapters that follow as illuminating as humanists.

I have chosen to start with the triggering epiphany of "On Being Polluted," emphasizing its 1976 provenance and all that it generated over the course of more than thirty-five years of publications and thinkings. The dates of the earliest published chapters reveal the chronology and development of prescient ideas that have gradually become naturalized as standard knowledge. Darwinian evolution, human choice, "free will," they all appear right from the start. Moreover, there is almost nothing in them that needs to be recanted or corrected. Today, I might refer to melamine in food imports from China instead of Kepone in waterways, but wanting to respect the contextual force of the originals I have kept my editing to a trivial minimum. The most recent chapters in this intellectual saga are the ones I have written especially for the book, inserted at the most opportune points. The conclusion, "My Life as a Robot," I hope will be experienced as inevitable and necessary, the sprouting of seeds unwittingly planted long ago. If the logic of this book counts for anything, it was inevitable for "me."

PART ONE / Ecology

CHAPTER ONE

Awakening to the "Environment"

T HE DISCOVERY OF "the environment"—in the contemporary sense—
is far from recent. From different angles of vision, books by Roderick Nash, Donald Worster, Max Oelschlaeger, and others have given us
historical accounts of this growing awareness, particularly as it acceler-
ated in the nineteenth century. And connections, however tentative, had
early been made between the environment and the human condition. A
literary type like me could not have failed to notice remarks about
weather and psychology from Samuel Johnson and James Boswell (see the
epigraph for chapter 4). George Eliot's need to escape London for the
countryside in order to relieve her headaches was a covert report on earlier forms of urban pollution. And E. M. Forster's *Howards End* concludes with a view from the country of the dark cloud of pollution hovering over London. One could go back to Juvenal and earlier if one
wanted to do a scholarly survey on this subject.

After the extraordinary advances in technology during World War II
and the resulting postwar rise in Western standards of living, more and
more people used more natural resources than ever before, with damage
to the environment growing more dramatically obvious. Since I will be
dealing here with the fairly sudden growth of my own awareness, the sixties are decisive. It was in the middle of that decade, sitting on the
screened-in porch of the low-tech farmhouse I was renting in the countryside within commuting distance of Detroit, when I began to notice articles in the *New York Times* about air pollution from industry and automobiles in seemingly distant parts of the country. Living as I did in almost
bucolic bliss except for drives into the city, I felt exempt and superior as
I read these early reports. How could I have felt otherwise? My farmhouse was situated on eighty acres surrounded by other eighty-acre farms,
the air seemed "pure," the countryside beautiful. I knew that Ford's River
Rouge plant west of Detroit was a source of smoke and smog, but that

was pretty far away from Rochester, Michigan, where I lived and taught, first at Oakland University and then, with a twenty-five-mile commute, as an assistant professor of English at Wayne State University. Although Detroit often looked smoggy, it was too early for me to connect the crushing headaches I used to have after a day at Wayne State with anything more than the admittedly stressful effect that teaching had upon me. At home in Rochester, though I felt mostly pretty good, I was beginning to experience a mild vertigo or dizziness one or two days per week, and it became troublesome enough for me to consider seeing my physician—until I realized the futility of trying to schedule a visit to take place on a day when I did not feel well. With my appointment at Wayne State coming to an end, a move was in the cards anyhow, and I decided simply to wait it out.

In 1968, after being offered a job in the English Department at Brooklyn College, I bought a ranch house on three-quarters of an acre on the north shore of Long Island in the village of Huntington Station. I had become enamored of the country and reluctant to give up the pleasures of vegetable gardening, so the long commute was a price I agreed to pay in order to have some land. A terrible mistake! The very first drive into Brooklyn was a shock, as I entered into bumper-to-bumper standstill traffic more than twenty-five miles from my destination. Though I had been born in New York City and had even been an undergraduate at Brooklyn College, years as a graduate student and professor in the wide-open spaces of the Midwest had alienated me from the New York ethos (so deftly captured by Saul Steinberg in his famous map of a New York–skewed United States) and the cramped Old World scale of the New York area's streets and roads. In my teens, Northern State Parkway, an idyllic, meandering engineering feat of the now much-reviled Robert Moses, had been a family pleasure drive on a beautiful Sunday afternoon at 35 mph, but now it was a dangerous, traffic-congested commuter horror with high speed limits that had never been envisioned as part of the original, multi-curved design. Much of its beauty had been destroyed by widenings that eliminated its parkway characteristics. The only alternative was the more recent and even more disastrous Long Island Expressway, already notorious as "the world's longest parking lot." From the moment of my entry onto the parkway, I could see the nimbus of pollution that seemed to hover way off in the urban distance and, curiously, to recede as I approached it.

The two years that I managed to survive the New York jungle were not wholly worthless, however, because without them I would not otherwise have met my future wife, Gloria, a fellow academic in Brooklyn's English Department. Her apartment on the Upper West Side of Manhattan startled me into remembrance of my New York childhood when I observed the thick particles of black soot on the windowsills and, even more startling, the black water in the bathtub after I washed my hair. For it was still the case in 1968 that domestic garbage was burned every day in incinerators accessed by a chute on each floor of most apartment buildings. It would be many years before I came to realize that the chronic respiratory ailments I suffered as a child, characterized by frequent and interminable sore throats, were the product of the city's intensely polluted air. Although it was not for that reason alone, I convinced her we had to leave New York and return to the Midwest, where we were lucky (or unlucky, as it would turn out) to find two jobs, hers at the University of Illinois at Chicago (then Chicago Circle) and mine at Indiana University Northwest, several miles south of Gary.

Gary! When I told my students at Brooklyn about the impending move, they thought I was out of my mind. And in ways they couldn't have known, they were entirely correct. I well knew the Dantesque nightmare they summoned up in their references to the pollution from the steel mills that no traveler on the Indiana Toll Road to Chicago could possibly have missed. I had passed through it many times during my graduate student years at the University of Wisconsin as I drove between Madison and New York. But I had no plans to be living in or very near to Gary. When Gloria and I drove out to Chicagoland to look for a place to move, I was determined to return to the life on a farm I had enjoyed outside Detroit. And after considerable exploration, we found exactly the sort of place we wanted. Located south of the Gary suburb of Merrillville in the Crown Point countryside more than twelve miles from the steel mills along Lake Michigan, five acres had been cut out of an eighty-acre farm that was still being plowed. Situated upon them, with majestic evergreens along the front, was a vintage white frame two-story farmhouse, to which an attached garage had been added in the rear that had already been converted into an independent apartment with its own kitchen, living room, bedroom, and bath, an arrangement that provided us with separate studies and space for scholarly work and reflection. The added wing could also double as guest quarters that would preserve our highly valued privacy. The grounds were lushly planted with flowers, gooseberries, mulberries, spring bulbs, bushes of all sorts surrounded by a considerable sweep of lawn indwelled by moles. On all sides of the property were large working farms that assured us of low population density and relative peace and quiet. The soil was perfect for a productive vegetable garden loved by possums and groundhogs that devoured our tasty crops. Corn borers and squash bugs were no less appreciative. But we got our share nevertheless—and the corn the raccoons left for us was a marvel.

The drive to the University of Illinois was about thirty-five miles and to Indiana University Northwest about twelve. Gloria had the worse end of the deal, especially in snow, but we felt it was worth it, even though the farmhouse required extensive rehabbing. During the first year the main kitchen was totally gutted and fitted with new cabinets and appliances, the main bathroom ditto, and a closet was turned into a new half bath. When the dust had cleared, after painting, insulation (the electric outlets had ice on them during the first year, the result of walls that were totally hollow), carpeting, wiring, plumbing, and whatnot, we were settled into what we thought was a perfect life.

Alas—we could not have been more mistaken. The growing awareness that we were inhabiting a polluted paradise involved a good deal of denial on our part after so much work, and my outbursts to colleagues and neighbors of my astonishment at the sheer density of pollution were met not only by denial on their part but by outright hostility. Pegged as troublemakers who failed to appreciate that the red skies over Gary were the local economic nutriment, we took on the stigmata of personae non gratae assaulting the goose that was laying golden eggs. (Years later, however, I would read over and over again about places like Gary whose lifelines—steel, coal, asbestos, uranium—turned out to be deathlines that were beginning to kill the workers who had formerly gloated over the benefits of their polluting jobs. As the delayed cancers began to take their toll, the outcry was always, "Why didn't they tell us?") But in their days of prosperity how sympathetic could the locals have been to those whining Cassandras and harpies from elsewhere who were threatening to take it all away?

The move to northwest Indiana turned out to be one of my life's radical cruxes. The repercussions were vast, not the least of them being the way it all fed into my ecological education, my own life's Big Bang, generating a new physical and psychological world for the rest of my life.

A great deal has changed environmentally since the early seventies, and this change is reflected in the chapters that follow. Written since the middle seventies, they are a record of the birth and maturation of the environmental movement over the past thirty years, while at the same time they form a kind of *bildungsroman* of my own ecological consciousness as it morphed into my consciousness of consciousness itself.

My ecological awakening began with the essay "On Being Polluted" (chapter 2), which found its way into the *Yale Review* in 1976, about three years after the traumatic events from which it issued in a white heat of astonishment. Rereading it more than thirty years later, I can relive the shock, the outrage, that caused it to be written. In those relatively uncharted waters of environmentalism in the early seventies, it dealt with phenomena that were not publicly discussed until years later: the extent of personal unawareness of the effects of pollution upon one's own behavior, mental acuity, emotional states, and physical performance; the need for an unobstructed vantage point (such as the one made possible for me by life on a farm) from which to notice changes in the weather and the resulting effects on air quality; the limited efficacy of standard monitoring procedures of air pollutants, skewed by their own technical and financial requirements. In 1973, when the essay was written, photochemical pollutants such as ozone were not yet in the news, and the stress was on sulfur dioxide and large particulates but not acid rain and the microparticles of diesel exhaust and from lawnmowers (and certainly not the effects of global warming). It was only then becoming apparent how extraordinarily far these pollutants traveled. New York and the New England states had not yet begun to complain about mutual pollution drift, let alone undertaken lawsuits against Virginia or Midwest power plants. Automobile and oil companies were resisting catalytic converters, gas mileage requirements, seat belts, air bags, collapsible steering columns, reformulations of gasoline—and building cars with recyclable components was a dream, though ultimately realizable in the future. Auto manufacturers, like tobacco corporations, were very far indeed from acknowledging the hazards of their products, which Ford so startlingly conceded in recent years with respect to its sport-utility vehicles, a striking sign, if one were needed, of the changed atmosphere, in all senses. Instead, they found it easier to persecute Ralph Nader for advocating for vehicle safety legislation. Still, the nuclear generating plant that Northern Indiana Public Service Company started to build along Lake Michigan was never to be completed, and the Shoreham nuclear plant on Long Island was finished but then abandoned. By the time "On Being Polluted" finally appeared in print, our own lives had changed as well.

The consequences of that inaugural cry of anguish and the experiences that produced it were manifold. To begin with, when the essay was published Gloria and I had already moved away from Crown Point, Indiana, to the farthest northwestern suburb of Chicago from which we felt we could manage to commute to our university jobs. The critical event had been a day of astounding smog that permeated the entire countryside. The neighboring farmhouses were almost completely invisible, the stench of chemicals from the steel mills was overpowering, and both of us were sick and depressed. Gloria had returned from the supermarket in Crown Point barely able to see her way back through the polluted fog. Once home, she stood opposite me in the kitchen as we looked out the viewless windows at the opacity surrounding us, and we decided on the spot that the time for evacuation had arrived. Gloria had finished the final draft of the biography she had been working on for many years, freeing us up for the move we were reluctant to face until then. This would not only entail giving up the farm and house so lovingly restored but would involve—years sooner than we could have expected—the sort of massive packing of books that most academics contemplate with dread.

And where would it be safe to live within the greater Chicago area? I had already learned at great cost that twelve miles counted as naught for the travel of pollutants. How far would it be necessary to distance ourselves from the steel mills, which in 1974 still cast a thick pall of smog as far north as the Chicago Loop, making nighttime driving on the Dan Ryan Expressway almost as surreal as the Indiana Toll Road?

The cost indeed had been great: our physical health was deteriorating, and our psyches had undergone a permanent alteration that changed our experience of daily life forever, adding categories of interpretation—of our bodily and psychological feelings and of the nature of the environment influencing them—that had not been part of our previous experience of day-to-day reality. Not only did I feel unwell most of the time, but during the four years in northwest Indiana I had reacquired a susceptibility to colds and respiratory ailments from which I had been freed after my exit from the sooty New York of the forties and fifties. Once again, sore throats were frequent and tenacious, requiring visits to the doctor for antibiotics (not available in my childhood), or else they would go on for weeks. For the first time since childhood I spent days in bed with colds serious enough to force me to cancel my classes, and I was beginning to realize, twenty-five years after the event, that my poor health as a child in New York City had been a casualty of the environment. The garbage we had thrown for years into the incinerator of our apartment house had been reassimilated as particulates by my vulnerable respiratory system. Nor was it encouraging to learn that the incidence of asthma and lung disease in northwest Indiana was considerably higher than normal for the country at large. How could I have leapt from the frying pan into the fire?

We ended up moving to the village of North Barrington, Illinois, about fifty miles northwest of the steel mills as the crow flies, seventy-five by car, giving me a pretty long commute back to Indiana University Northwest. This was as far as we could manage to go but, as later chapters will reveal, it was not far enough, a fact that I recognized soon after we bought our new house but had not yet moved in. During some of the preliminary drives to North Barrington from Crown Point and during some of the time spent there fixing up the house before we moved in, it was impossible to deny that I was already experiencing many of the same symptoms that had driven me out of Indiana. Although the relocation did in fact produce a marked improvement—my respiratory problems immediately ceased, and I felt free of air pollution symptoms for a larger number of days per year—the physical malaise and the alterations of consciousness that had become a familiar concomitant of bad air did not go away. They simply changed character and frequency as their relation to the seasons, wind directions, temperature inversions, and chemical mix carried by air currents were altered by a change of position.

My professional life, as well as my everyday life, was also undergoing a change, although I could not have been aware of it at the time. Moving to northwest Indiana and writing "On Being Polluted" constituted a major road selected and innumerable others rejected. The *New York Times* asked me to do a distillation of the essay for its Op-Ed page, which appeared as "Life in a Vacuum Cleaner Bag" on November 7, 1976. Although I was not altogether pleased with the title they gave it, which reduced a complex problem of air pollution to mere dirt, I was nonetheless glad to have it widely disseminated. It was reprinted by a number of newspapers around the country and brought me a small pile of letters in reply. One of them, from a thoughtful woman in the Midwest who saw the issues in religious and philosophical terms, planted a seed in my consciousness that grew for months until it exploded not long after as "From Transcendence to Obsolescence: A Route Map," which appeared in the *Georgia Review* in 1978 and reappears as chapter 3 of this book.

If "On Being Polluted" was a specimen of practical environmentalism, in which I addressed concrete material circumstances connected with daily life, "From Transcendence to Obsolescence" was a ruminative, reflective, philosophical consideration of those same circumstances as seen from a more cosmic distance. Its point of view was well suited to the seventies, and much of what it had to say is still appropriate today. But things have undoubtedly changed, partly as a result of the environmental movement itself and its effect on public and private awareness. This essay, literary, allusive, always cognizant of the Western tradition, was also to play a role in the emergence of ecocriticism as a discipline and in the transformation of my professional interests, eventually embracing evolutionary and consciousness studies, another one of those unforeseen consequences.

After the publication of these pieces, key administrators on my campus turned hostile, although I was never called upon the carpet or treated as a whistle-blower. Gary, after all, was a company town, and U.S. Steel a benefactor of the university. That I was not merely paranoid was made pretty clear by a surprising telephone call I received from a student I had never met. She phoned to inform me that at a community meeting orchestrated by U.S. Steel, copies of my ecological writings had been handed out, discussed, and censured. Somehow, these forces of the community seemed to be saying, I had let the cat out of the bag: it was as if they believed that without my efforts, the world would never have noticed that Gary, Indiana, was a maelstrom of pollution.

It was not until we moved to the northwest suburbs of Chicago that I had amassed enough experience to write "Air and Being: The Psyche-delics of Pollution," now presented as chapter 4. This essay continues as a unique account of the subjective realities of life in a polluted place and as a seminal precursor to the consciousness-oriented chapters that conclude this book.

CHAPTER TWO

On Being Polluted

DESPITE ALMOST DAILY articles about pollution in the newspapers, the nature of life in a polluted environment remains for most people rather vague. The *New York Times,* for example, has had a series of startling articles about air pollution in Japan, in Riverside, California, and in Los Angeles, but my impression is that people, even those living in polluted areas, are not quite sure what the effects of it all are supposed to be. It is true that in Los Angeles people complain about their burning eyes, and in New York and London some people complain about the stench of automobile exhaust, but that is about the extent of it. One hears on television about distant unfortunates in obscure places like South Holland, Illinois, where diseases of the lungs are much more numerous than in the rest of the country, but that is explained as the result of South Holland's dreadful location in the midst of oil refineries and steel mills south of Chicago, and what is understood is that such a situation is relatively rare and special.

What I would like to convey, solely as a layman, without official data, figures, or tables, is the actual experience of daily life in a very polluted area, an area which has many of the characteristics of urban and suburban areas all over the United States and which, far from being small and local, spreads over hundreds of miles. The extent to which pollution affects one's lifestyle and one's health can be startlingly great, and I think that most people are unaware of the effects air pollution may be having upon them.

In the spring of 1970, my wife and I spent a week or so looking for a small farm within commuting distance of Chicago and Gary, but one which would be beyond suburbia and far from the madding crowd. After considerable search, we found almost exactly what we wanted: four and a half acres with a large and authentic farmhouse, ravishingly beautiful grounds, and lots of trees and outbuildings. The whole little farm is surrounded on all four sides by fields of corn and soybeans. About five miles southwest lies the charming city of Crown Point, the county seat of Lake County, Indiana; and about fifteen miles to the north is the city of Gary, seemingly far enough in the distance not to bother us. In one year we had almost completely renovated the farmhouse, inside and out, and during the second we tackled the outbuildings, the driveway, and began to work on the landscape. During the first year we were always very busy, putting in fourteen-hour days quite regularly, painting, carpentering, and so forth. The weather was mostly good, although several months of extreme wind during the early stages made us feel that perhaps we had settled in an uncongenial climate, for the first fall and winter were plagued by fierce winds from the west. Eventually, however, the winds became less regular and, as a result, less vexing. Except for an early morning smell of "tar," which I at first attributed to road repairs, we had few signs of anything that might be considered pollution, except for the characteristic red and smoky skies over Gary, fifteen miles to the north. But after many mornings of "tar," it began to hit me that the smell, which usually dissipated around noon on warm days, was probably wafting in from the industrial areas around Lake Michigan, and I eventually learned that the odor which I referred to as "tar" was from oil refineries near Gary.

The pollution in Gary itself was always too obvious to be missed, even from a distance. The first time I ever caught sight of it was years ago when I was a graduate student at the University of Wisconsin. The trip between Madison and the East involved passing right by the steel mills on the Indiana Toll Road, which struck me then and strikes me now, no matter how often I go through it, as a nightmare out of Dante. What I was not prepared for two years ago, when I first explored the Chicago area before moving here, was the dense, smoky pollution south of the city, of which a vivid view and smell are available to the driver on I-94, the Dan Ryan Expressway. Once we settled on our farm in Indiana, the drive to and from Chicago became quite an experience, one which I came to term, after the title of an obscure Renaissance poem, a "banquet of sense." Every few miles has its own distinctive smell, its own ambience: one moves along from the "baked potato" smell to the "chives" smell to the smell of burning garbage from the city dumps south of Chicago, to the Sherwin-Williams orgy of sights and smells, a paint factory that looks as though it came from, and should return to, outer space. After a trip home from Chicago, one emerges from the car a bit woozy, one's senses raped. But it is necessary to concede that the trips to and from Chicago, although one may weave a bit while driving, are made bearable by excellent roads, at least in contrast with the traffic horrors of New York, and also by the fact that Chicago is worth getting to, since Lake Shore Drive and the Chicago Symphony are two of the world's great pleasures.

During the winter of our first year on the farm, the pollution began to be more in evidence, not only south of Chicago and throughout the Gary area, but also, to our nervous dismay, quite pronounced and smoky around our house. But since we could close the door and cuddle up in our comfortable living room, perhaps in front of the fire, and since most of the days were reasonably decent, we reluctantly accepted the smoky nights as the price we had to pay. And the spring, summer, and fall of 1971 were absolutely splendid, except for a few shocking days in late spring. On one of these, which I remember as May 8, we decided to sit outdoors with friends of ours who had come from Detroit to stay with us for the weekend in order to hear Solti conduct Mahler's *Eighth*. We emerged from the house into an atmosphere of warm, murky sunlight, obscured by thick brown fogs that were blowing in from the north and sidling around and between us as if we were flying at ten thousand feet through the clouds. It was a full-scale pollution attack from Gary and the horrible north and it was poisonous—our eyes began to burn, our noses rebelled at the stench; one had a sense of being cheated: a beautiful day had been stolen by others and they were pouring their garbage into it and into us. And only a few miles away, someone else *was* having a beautiful day—until the winds might shift and *he* became the victim. It was a terrible shock.

But a beautiful summer ensued, and a marvelous fall, made possible to a large extent by a partial closing down of the steel mills because of an earlier overproduction (happily during months when the south wind blows the pollution over Lake Michigan) in anticipation of a strike that never took place. The skies over Gary, which we can always see quite clearly, were unusually clear. The *New York Times* even ran a story on the oddity of clear skies over Gary. Indeed, the familiar red haze was gone and the whole atmosphere, as far as we could see on all sides, was bright and crisp. It was a novelty to be able to look northward and see a sky as clear and free from red smoke as the sky in the south. This interest in the weather had been with us from the first day we moved here, because our location provides a view in all directions that is unimpeded by anything much taller than a row of trees on a little hill. We can see far into the distance and have a good idea of what the weather will be like later in the day, because we can see it for miles and miles.

By the beginning of 1972, the situation changed dramatically. The intermittent pollution, which had been bearable though very unpleasant, was suddenly transformed into daily or almost daily full-scale invasions from the north and west. Almost every day in January we awoke to find ourselves out of sorts, with severe burning eyes, curious oppressive headaches, and a general feeling of strangeness. This took place four or five days a week for perhaps two months. And on each of these days upon arising we could look out of our bedroom window to the north to see the gray, murky sky spilling over upon us from Gary and East Chicago and floating onto miles and miles of land to the south of us. Although all of this was bad enough, by late February things changed for the worse. The burning eyes gradually diminished, but we would arise in the morning to experience a congeries of other, more distressing ailments: dizziness, nausea, tingling pains in the extremities, and a dazed, lethargic aimlessness. I would sit at breakfast, hovering and swaying over my coffee, with a sour and rumbling stomach and a desire to throw up. Sometimes this would last for most of the day, day after day for weeks, with only a few intermissions. It was usually worse if we went outdoors. And on very severe days, when the fog of pollution was so strong that we could barely make out the house and grounds of our neighbors about a sixteenth of a mile down the road, we would begin to fall into a depression at having to go through another day of it. Once or twice, on horribly polluted days, we went for a drive, hoping to find the outer reaches of the pollution, to find a few breaths of pure air, and (though we hated to admit it to ourselves after over a year of strenuous work on our house) perhaps a safe place to move to that would still be within commuting distance of Chicago. But even though we would drive another twenty miles to the south and east, farther and farther from Gary and Hammond and East Chicago, we remained in a fog of stench just as bad as it was around our own place. And the physical symptoms persisted.

Since our ability to read, to concentrate, and to get things done was often impaired while we stared bovinely at the walls, I decided to visit our doctor for general examinations. These indicated that I was in perfect health, but when the symptoms persisted, I then went back to the doctor for blood tests and an electrocardiogram, for eventually chest pains began to enter the picture. Once again I was found to be fine, except for the symptoms, which the doctor agreed were most likely the result of pollution. He gave me some pills to take for burning eyes and related phenomena, but, like most antihistamines, they produced their own unpleasant symptoms. My wife's principal complaints were headaches and burning eyes, but she too was found to be in satisfactory shape by the eye doctor she went to see. The only thing that remained certain—and still remains certain—is that the symptoms appear only when the air is polluted.

The effects of all this on our daily lives became greater all the time. The correlation between north and northwest winds and our maladies was perfectly dependable, and as a result the first thing we would do on arising and feeling wretched was to look out of our bedroom window to the north. At such times, there was invariably pollution pouring upon us from the north, and if we arose feeling well, there was equally invariably an east, west, or south wind. As the northerly component of a west wind increased, so did the pollution and our sense of feeling ill. Another daily ritual had become the checking of the local newspaper every afternoon to see what the prevailing wind was expected to be and to see what the pollution readings were for the previous day. Our lives had become geared to the winds—and with good reason. Sometimes a beautiful day would suddenly change into something out of a bad dream because of a shift in the wind. Sudden temperature inversions are a frequent phenomenon in northwest Indiana, and when they occur, a clear day with temperatures in the upper seventies can change into a thick and almost opaque fog with temperatures in the forties, all in the space of half an hour. Opening and closing windows, even on more sedate days, is a futile and dangerous activity which we eventually learned to avoid. I recall one shockingly polluted day when I was working out in the garden, dizzy and depressed, waiting for my wife to return from Chicago. Suddenly, around five in the afternoon, the wind shifted and within twenty minutes the sky became clear and the fog and smells completely dispersed. I raced into the house to find that it reeked like the Toll Road in Gary, for when the weather clears after several hours of pollution, the house has already filled up with the smelly, smoky air and retains it dramatically for several more hours. I threw open all the windows to air out the house with the newly arrived fresh air, just in time for my wife to return from a harrowing ride through the baked-potato smell, the chives smell, the dump smell, the Sherwin-Williams smell, and the Gary stink to our now sweet-smelling house. And just as she entered, the winds shifted again and within a moment, pressing in from the north was the whole sea of garbage about to engulf us for another round. I quickly closed all of the windows to preserve a few extra hours of grace, and within half an hour we were once again in the heart of the heart of pollution.

After several months of this kind of life, feeling rotten four or five days a week and depressed and anxious on the other two or three, we decided that we couldn't go on in this fashion for much longer, that as much as we were attached to our farm, an island of beauty in a floating sea of garbage, we would have to do something or move. We both agreed that it would be worth almost any price to try to keep the farm and make it livable, and so we began to investigate the potentials of adding electronic air cleaning to our heating system.

With a view toward obtaining information, we spoke with our heating man and were told that electronic filters were able to remove sulfur dioxide from the air as well as all airborne particulates, including dust and pollen, and the literature I read on the subject (admittedly produced by manufacturers) claimed a 90 to 99 percent removal of these pollutants. To confirm this, I spoke to my doctor, who agreed that such a filter might help the situation, although he did not think we should by any means expect a miraculous elimination of the total problem. In an attempt to obtain further and more precise information, I phoned the Lake County Health Department in Crown Point to see if they had done any testing of electronic air filters or had any government reports on their efficiency. I was put in touch with the man regarded as the pollution specialist of the department, and I asked him what he knew about the effectiveness of these filters on local air pollution. His reply is inscribed in my memory cells: "What pollution? There is no pollution in the southern parts of Lake County." There was a very long pause, while I felt both panic and blankness of the mind. When I was able to speak, I began to tell him the whole story of our problems while he became more and more belligerent. He informed me that the department had done a pollution test and published a report showing that the pollution outside of Gary and the other industrial cities on Lake Michigan was nil, that pollution in Crown Point was almost out of the question. To my incredulous question, "Then what is all the black garbage that floats and stinks over this whole area?" his only reply was, "Have your doctor get in touch with us if you are being treated. So far, we have never had a single complaint about air pollution and, as for your symptoms, I never heard of such symptoms from air pollution."

It was about a month before I could get up enough steam to make an actual visit to the Lake County Health Department to confront this man in person, but during the interim I spoke to many people about air pollution, not, admittedly, for the first time, but now as an obsessive subject. And the result was quite shocking, for I discovered that most people, for one reason or another, were unaware most of the time of any actual pollution, even though they were quite aware intellectually of the existence of a problem. This apparent insensitivity to a stimulus that was ruining our lives seemed to be traceable to three main causes. The first of these was simply the fact of a person's being a native (or a long-term resident) of the Chicago-Calumet area. Most of my students at Indiana University Northwest (located in a suburb of Gary) fell into this category. The IUN campus is located about three miles south of the steel mills and other industries on Lake Michigan in a pleasant residential area that often is clear on badly polluted days. But when the wind is right, on a considerable number of days of the year the classrooms are permeated with a stench from pollution that is unbelievable. I have actually driven to class on polluted days to arrive in a daze, on the verge of throwing up from nausea, have walked into class where the smoke was almost palpably visible, to say in amazement, "The pollution is incredible today," only to have a student reply, "What pollution?" (whereas visiting friends of mine from other parts of the country have expressed astonishment at pollution half as bad). This repeated experience gradually made it clear, although it was hard to fully believe it, that living in these smokes and smells for all of one's life tends to make them part of the normal, unnoticed environment.

A second cause of the obliviousness of pollution I found to be the fact that most of my friends and my wife's friends are academics from the Chicago-Gary area who live in the northern suburbs of Chicago where there is often little pollution. Even in Chicago itself, from the Loop northward, the pollution is mostly from automobiles and only on certain days do the winds carry really bad pollutants into the city. This phenomenon is the result of the prevailing winds in the region of southern Lake Michigan. Most of the year, especially winter and spring, the winds are west or north and once in a while east. During these periods the northern suburbs and the north side of Chicago escape the pollution, while the entire region south of Chicago, from Joliet to Valparaiso, is steeped in pollution of the worst sort. But during the summer, when the winds are usually southern, residents of northwest Indiana experience a good many beautiful days, even in Gary itself, while the pollution is blown over the lake or, if the wind is southeasterly, over Chicago and its more northern areas and suburbs. But most of the time, if you live in Evanston or Skokie, you are in a situation not at all comparable to that south of Chicago. My academic friends, then, are not much acquainted with pollution, unless they happen to live in Hammond or other cities in northern Indiana. And in those few cases, they have usually lived there a long time and do not particularly notice the pollution.

The third cause of unawareness that I have observed is the smoking of cigarettes and other tobacco. I do not think I have come across a smoker yet who is affected by air pollution, that is, consciously affected. Smokers live in an ambience of smoke and smells through which not very much can penetrate from the outside world. One is almost tempted to conclude that smokers are in the advantageous position, vis-à-vis pollution, of doing themselves in pleasantly, instead of being done in involuntarily by industrial smoke and gases. Of course, one has read in the papers that smokers may be doing *other* people in by polluting the air indoors for nonsmokers. Is smoking a ruggedly individual way of telling United States Steel "you'll never catch me alive"?

These experiences with so many people unaware of air pollution do not increase one's sense of being in touch with reality. And so, a bit desperately, although we never for a moment felt any correlation, we decided to follow the Lake County Health Department's recommendation that we have our well water and our furnace tested as possible sources of our ailments. As it turned out, there was nothing wrong with either of them and, in any case, our symptoms appear only on polluted days, whether we have had water to drink or not or have been in our house with the heat on or not. Indeed, some of the worst attacks have been on mild spring days when we were away from the house. Needing further assurance, however, I finally wrote two long letters to William Ruckelshaus of the

Environmental Protection Agency (EPA), the second of which was written after I had got up the nerve to visit the Lake County Health Department for an encounter with the "What pollution?" man.

As things turned out, the day I went to see him was badly polluted, and I drove into Crown Point under dark and gaseous skies. My conversation with this employee provided a coda of Kafkaesque unreality, for as he reported on the data that the department had put together in a booklet, I sat glassy-eyed and dazed in a chair opposite him, a bit sick to my stomach, while he told me once again that there was no significant pollution in the southern portions of Lake County. For me, the high point in our conversation was reached when he reiterated his unawareness of symptoms caused by air pollution when, only three minutes before, while waiting to see him, I had observed on the wall of the office a large poster from the EPA dealing with the effects of air pollution on plants and people. Weaving and a bit dull, I had just read, with a mixture of relief and fright, that standard effects of air pollution on humans were dizziness, nausea, pains in the extremities, headaches, burning eyes, lethargy, etc., etc.

Before leaving, I requested a copy of the Lake County Health Department's Report on Pollution, which I eagerly read as soon as I reached home. The health agent's description of the report was more or less accurate (although there was more pollution indicated than he was willing to concede) but the report itself could hardly have been more misleading. On the basis of tests performed on random days during a nine-month period of one year, the department had come to the conclusion that air pollution in Lake County was not a serious problem outside of the immediate industrial areas. What was so amazing about this report was that its random data were collected during the portion of the year when air pollution is at its minimum in Lake County and that it was based on one year only. Even worse, it went against obvious daily experience of the senses. As for the span of the test, I myself had seen from just two years of living in Lake County that the summer and fall can be very fine if south winds prevail, and I had also seen in that same short period that one year can be entirely different from another, with respect to winds, temperature, rainfall, and, most important, pollution. And so, on the basis of this nonsensically flimsy test, done up in "scientific" regalia, the health department had taken its fantastic and dangerous position that air pollution was not a serious problem here, while thick clouds of every pollutant under the sun were passing through their very noses.

In response to my letters to the EPA, I received a very detailed and sympathetic reply from Ruckelshaus, treating the pollution here as an obvious fact of life and outlining measures that are now being taken to remedy it. Bizarre as daily life can be in "Chicagoland," it seems to be the case that action is beginning to be taken, although remedies are a long-term affair. The city of Gary, as well as the EPA, have finally succeeded in convincing United States Steel of the urgency of the problem, and a five-year plan for rebuilding coke ovens, a principal source of sulfur dioxide, is claimed to be under way. And the Gary newspaper, the *Post-Tribune*, despite its delicate position in relation to the steel mills and the workers, has consistently taken a firm stand for reform, at least during the year or so I have been a reader. These things are encouraging, although I doubt whether my wife and I can contrive to hold our breath for five years. For the moment, we have spent about six hundred dollars to install electronic air cleaning equipment in our forced-air heating system.[1]

I would now like to do what may cause many people to look askance, namely, to draw some conclusions entirely on the basis of my own personal experience and the testimony of my senses. This is probably an unscientific thing to do but, on the other hand, it would be hypocritical of me to pretend that I do not regard the reports of my senses as reliable descriptions of reality, at least of reality as lived by people rather than machines or testing instruments. As one instance among many of the reports of my senses, I must recall the morning on which I awakened to feel all of the familiar symptoms of pollution only to find, upon looking out of the north-facing bedroom window, a beautifully clear sky as far to the north as I could see, which means to Gary and Lake Michigan more than fifteen miles away. I was baffled. Could it be the case after all that I am simply a psychosomatic nut? Happily, or unhappily, depending on one's point of view, it turned out that I went to get the mail from our roadside mailbox only to discover that a northeast wind was blowing everything to the west. As a result, the sky immediately to the west of our house (and beyond the range of the bedroom window) was heavy with thick, smoky pollution, leaving the north, which was the source of it all, crystal clear to the east. Later that day, my wife and I went shopping in nearby Merrillville (a bedroom community that fancies it has "escaped" from Gary) northwest of our farm, right into the pollution's heart, and we both experienced one of the worst attacks we had ever had, so bad that my wife, who gets only some of the symptoms that I do, this time had them all. My driving was so impaired that I could gauge only with difficulty the requirements of the road and the movements of other cars, and I found it hard to react quickly with necessary maneuvers. Both of us were utterly dizzy, nauseated, and rapidly developing tingling pains in our arms and legs. We got home as fast as we could and literally sank onto the sofa, where we sat for an hour staring stupidly at the walls. The next day, we took a preliminary reconnoiter to Kankakee, Illinois, to get the lay of the land as a possible future home after a thorough pollution check (although we do not expect to write a letter of inquiry to the Kankakee Department of Health).

The first and major conclusion I would like to draw is that the lives of millions of people, whether they are aware of it or not, are deeply affected, physically and psychologically, by air pollution. Their daily moods, their sense of the way they feel each day, even the events of their lives, are greatly determined by the nature of the air. Let me provide an illustration.

It is generally well known in these parts that auto accidents on the Indiana Toll Road increase on heavily polluted days. There is a considerable stretch of this road that runs right by the steel mills and other sources of pollution, and on certain days the visibility is close to zero. Not only is this the case on humid, foggy days, when the smoke pours out of the chimneys of countless factories right down onto the road, but even on days that are clear elsewhere in the area the road can be obscured by smoke if conditions are right. It is common to find newspaper reports of accidents and deaths on the Toll Road alluding to the role of smog from industry as a causative factor. But I would like to carry this much further by submitting that pollution of the sort that is floating around big cities, suburbs, and rural areas nearby, and which is present many or most of the days of the year, causes lethargy, vacancy, wandering of attention, lassitude, faulty focus—in a word, it causes something very akin to drunkenness. The pollutants enter the bloodstream and perform like alcohol and drugs, causing (among other things) all sorts of accidents. But this effect differs from the effects of alcohol or fog in that people are not aware of its connection with a given cause and as a result are not vigilant about it. One knows to exert extra care if one is a bit high or if one cannot see clearly, but if one is falling into a subtle (or not so subtle) torpor and the environment is fading from view, very undramatically and seemingly uncaused, one is simply unaware of what is happening to oneself. One can report to the police that fog on the Toll Road has caused one to crash into the rear of another car, but how many people are likely even to think of thinking that the polluted air has impaired their senses and their consciousness in some other way?

There is a good reason for this unawareness. Most people live in cities or towns that make it impossible for them to see the weather. If one is traveling into New York City or Chicago or Los Angeles from the suburbs, one does indeed see that one is entering a haze. But if one lives in the haze or spends most of the day there, and there is no point of reference because no other place is visible except where one is, how is a person to obtain a basis for awareness? To connect the quality of the air with the quality of one's day-to-day feelings, one must be able to perceive the air. This has been shown to me very forcibly by my own life on a farm with unobstructed views of all four directions. One emerges from the house to a more or less clear day. One looks to the north and sees a polluted sky. A wind starts. And what was a clear sky nearby begins to be overtaken by the red and gray sky in the distance. As it gets closer and closer there comes a moment when a smell suddenly starts. I have stood in the fields and said to myself, "It will smell in thirty seconds," and it *did* smell in about thirty seconds.

Or take another day, like the one we had recently, which seemed to be perfectly clear but on which I nevertheless felt absolutely rotten. I looked to the north and saw the familiar red sky of Gary. I looked above our house to see a beautiful, clear blue sky. There was very little wind (but it was northerly) and it looked like a totally gorgeous day. But then I looked to the west, to the east, and to the south, and what did I see? A red haze on all sides, a halo of red, very deep in the north, gradually becoming pink as it ringed the east and west, and lightest pink of all in the south. And I ask myself a simple question: how can it be a beautiful, clear, unpolluted day above my farm if there is pollution visible on all four sides around me? Answer: it is not a beautiful, clear day. I am in the midst of a sky that is polluted for at least fifty miles. And that is why I feel miserable, why my stomach is sour and rumbling, and why I am compulsively eating sweets all day long. It is a horribly polluted day and I am suffering from symptom X: "pollution stomach." (The next day, the Gary newspaper gave high sulfur dioxide readings for the previous day.)

To make connections between weather and one's physical state one needs to live outside of the city, where the visibility is excellent. How much more difficult it must be to be able to connect one's mental state with the effect of the weather on one's body in urban areas hardly needs to be pressed. But if one is feeling physically out of sorts, if one is unable to focus, feels lazy and vaguely wrong, one's frame of mind is altered, and one's interest in things is not what it would otherwise be. It is a cliché nowadays that the current crop of young people are aimless, passive, and inattentive, and much of it is attributed to the mushy-mindedness of television. But is it inconceivable that a whole generation of urban and suburban youth might be, if I may use an extravagant expression, permanently stoned?

> I taste a liquor never brewed,
> From tankards scooped in pearl;
> Not all the vats upon the Rhine
> Yield such an alcohol.
> Inebriate of air am I,
> And debauchee of dew,
> Reeling, through endless summer days,
> From inns of molten blue.[2]

I won't press it—but the *New York Times* article on air pollution in Japan pointed out that schoolchildren in different sections of a very polluted industrial city had different personalities, depending on which pollutant prevailed on their side of town. Why not here too? Why not a sulfur-dioxide personality? The problem may really be that "I *don't* taste a liquor never brewed."

If my suggestion regarding ignorance of causes has any validity at all, would this not make the problem of air pollution an even greater one, beyond reckoning, than it is even now considered to be? Automobile accidents and deaths, vague illnesses and pains, trips to physicians and eye doctors, perhaps even family squabbles and difficult children—is it beyond all reasonable seriousness to suppose that millions of people's lives are influenced at every turn by a "drug" whose very existence they are scarcely even aware of, despite all the hullabaloo in the newspapers and on television? Is it just those "cardiac and respiratory patients" whom we always hear about who are in peril?

For the whole business of pollution watches and warnings is very misleading, with the implication that people who are in precarious health are the only ones in danger. Why is it not actually the case that everyone is in danger, not only at these dramatic moments, but most of the time? When one considers the weird solicitude about people being damaged by a few marijuana joints, while the major part of the population is being force-fed "smoke" like geese in a goose-liver-paste factory, one no longer knows what it means to make sane judgments. And the pollution readings carried by daily newspapers can also be misleading, despite their good intentions. The Gary newspaper carries a pollution report for downtown Gary on the front page almost every day and, in general, I find that the readings tend to correspond with what I see and feel, although it is often the case that areas at considerable distance from Gary have severe pollution when the Gary readings are low, because certain weather and wind conditions seem to carry and then dump the pollutants in faraway places. But to many people these daily reports must be confusing. According to the Gary *Post-Tribune*, a sulfur dioxide reading of .11 is to be regarded as the "danger level." So far, however, the highest reading I can recall was roughly .055, considerably less than the danger level. Yet those debilitating days when we have felt extremely ill, when driving and concentrating were difficult, received readings of perhaps .033. A sulfur dioxide reading of .11 strikes me as all but inconceivable. That is, there might be a few hardy people left around the next day to collect bodies, but it is hard to believe that .11 is merely "dangerous." As for pollution warnings, they would appear to be a kind of bread and circuses that encourages quietism and obliviousness.

This quietism is reinforced by two folk myths that are to be heard everywhere about remedies for pollution: air-conditioning and staying indoors. Certainly these two devices are not entirely useless, for when the air is thick with pollutants it is better to be indoors instead of having it all blown in your face, but the benefits are much exaggerated. For what never seems to occur to most people is that there is only one source of air: the outside air. The air in their houses is outside air that has come inside. No matter how tight the construction of the house may be, or how excellent the storm windows, it is not possible to prevent outside air from entering. *The inside air is outside air that has entered.* If no air were to enter, one could not survive. All that can be altered is the rate of entry. Exhaust fans, furnaces, fireplaces, gas cooking and water heating, and ordinary breathing, all of these use up oxygen, which is replaced by new air entering the house. The new air enters partly because of the moving air outdoors and partly because it is sucked into the house as the internal air is used up. Thus, no amount of staying indoors can do more than slightly reduce the effects of pollution. Finally, after one hour or four hours or however many hours, depending on the construction of the house and what is going on inside and out, all of the air inside has been completely changed. It may take longer for polluted air to enter a tightly constructed house, but it also takes longer for it to leave when the outside air has improved. And, of course, air-conditioning merely cools and dehumidifies the air that is available. It does nothing else. The filter on the average air-conditioner can barely remove large dust particles. It certainly is incapable of removing anything as small as pollen or air pollutants.[3] Again, it is true that cooler, drier air is easier to breathe, and one feels better, even with hay fever, breathing such air, but since nothing much besides water has been taken out, air-conditioning provides minimal evasion of the bad composition of outside air.

Electronic air cleaners are said to be capable of taking most dust and pollen and pollutants from the air. In our own limited experience we have found no relief whatever from hay fever since the installation of air cleaning, although we have found that our symptoms from industrial air pollution have been substantially reduced during pollution attacks. Our house has a modern addition which has its own heating system, independent of the main house and thus without the benefits of the air cleaner that is in the main heating system. The differences in our condition are easily observable in relation to which part of the house we are in during bad pollution spells. We recently bought a floor-model air cleaner for the modern wing, though I might point out that the smell of ozone (which these air cleaners produce) is quite strong with a floor-model unit, whereas the smell is so evenly dispersed throughout the main house by the central air-cleaner that it is barely detectable there. In any case, our electric bill has risen dramatically, the main air cleaner adding somewhere between ten and fifteen dollars a month to the bill, since the furnace blower and air cleaner run twenty-four hours a day. All of this, of course, adds to the pollution problem by consuming more electricity. In our region, Nipsco, the power company, is one of the chief polluters!

Finally, one is left with the problem that, as things stand, there is little one can do to avoid pollution except to live in a place where the air is still pure (wherever that may be). It is sadly true that living in the country, even on a farm, is not necessarily a solution, since the travel of pollutants is very far indeed. The recent environmental conferences in Sweden yielded a report that sulfur dioxide can be carried five hundred miles with ease.

I think, however, that a major turning point would be achieved, beyond what is now being done, if the average person were to realize the extent to which his health, moods, even his life itself, may be influenced by air pollution. What is urgently needed is a large-scale survey to be administered by the government (or a private agency) to urban and suburban populations to determine what in fact is the relation between people's day-to-day physical and mental states and the quality of the available air. My own feelings are that the information obtained through such a study would be devastating.

CHAPTER THREE

From Transcendence to Obsolescence

A Route Map

ALTHOUGH THE AGE-OLD problem of the conflict between body and mind that has tortured philosophers from Plato to Kant and obsessed the Church from Augustine to Pope Paul has been resolved in modern philosophical thinking by the elimination of "mind" as an autonomous entity, the conflict would appear to have returned again to haunt us in a new guise. The idealized emphasis on "rational" in the concept of man as the rational animal that characterized Platonic-Christian thought for two millennia had generally been the product of man's sense of his own physical weakness, his knowledge that Nature could not be tamed or bent to his own will. In lieu of the ability to mold Nature to serve his own ends, man had chosen to extol and mythify that side of his being that seemed to transcend Nature by inhabiting universes of thought that Nature could not naysay. The triumphs of intellect and imagination by thinkers and artists, and the heroic transcending of the body by saints and martyrs who said "no" to their earthborn limitations, provided for centuries the consolations of a victory that could be obtained not by winning the battle but by changing the battlegrounds.

In the course of human history until the twentieth century, there was never any serious likelihood that the body-mind battle could be won on the field of the body. If one found that it was necessary to produce ten children in order to insure the survival of five, if one could be swept away by plagues that killed hundreds of thousands, if one lost one's teeth by thirty, could not be certain of a food supply for more than a few days, carted one's own excrements out to the fields or emptied chamberpots out the window, one could hardly come to believe (despite humanity's fantastic ability to believe almost anything) that one's ideal self would ever stand forth on the field of the body, in the natural world. Nature was indeed the enemy, whom one propitiated in the forms of gods and goddesses or saints and martyrs, but who would finally do one in en route to one's *true* home, Abraham's bosom. Good sense taught that it was pointless to waste what little life one did have in a quarrel with the cruelty of Nature when the rational solution could only have been to accept a final repose in the kindness of God. If humans were indeed made in the image of God, then it was reasonable to assume that only God could fully appreciate "man's unconquerable mind,"[1] while a just assessment of reality required that the field of the body be given up—as how could one do otherwise?— to Nature.

The exaltation of religious figures during all of Hebrew-Christian history prior to modern times was an acknowledgement that saints, prophets, priests, and nuns more fully embodied spiritual ideals than did most people and that an approximation to spiritual perfection, however difficult, was a more realistic goal than that of bodily self-sufficiency or domination over Nature. The fascination with the fall of heroes in history and fiction involved a painful recognition that nothing physical could endure, not merely in the obvious sense that everything created must inevitably die but in that everything created can barely stay alive. The philosophy of *carpe diem*—make your sun run fast if you can't make it stand still, to echo Andrew Marvell—was never a prevailing one. For most people, the fear of human fragility and a lack of substantial power against the material world made profound self-confidence a luxury only for kings, who themselves derived their power from God. For others, realism required an acceptance of the Divine will: existence was a gift, and the creature had no rights. All was grace.

But by the eighteenth century, the rise of industrialism in the West was accompanied by a decline of religion that cannot be seen as an accidental concurrence. And from then on the trend accelerates. As the average person becomes more enabled to live in comfortable houses that resist the elements, to escape most of the childhood diseases that had made fecundity a virtue, to preserve his teeth into middle or old age, to store food for weeks, months, or years ahead, to communicate rapidly through time and space, to move long distances with ease, to dispose of excrements through indoor plumbing that makes them all magically vanish in a trice, his perception of Nature undergoes a startling alteration. No longer does Nature seem quite so red in tooth and claw; for a person is much less likely now to perish from the heat or cold, to starve for want of food; his formerly intolerable dependency on the caprices of Nature is no longer so gross; his relation to the other animals and to the vegetable creation appears thickly veiled—by air conditioning, frozen foods, washing machines, detergents, automobiles, electric blankets, and power lawnmowers. And most startling of all, his need for transcendence seems to fade away. For what, after all, is so dreadfully unpleasant about contemporary Western middle-class life that it needs to be transcended? Yes, of course, traffic jams on the freeways are a strain and suburban life can be parodied, but on the scale of things, in relation to our historical life on earth, the ills of suburbia are not so drastic as to encourage an unduly hasty shuffling off of this mortal coil.

It has been said again and again that modern Western society's comfortable life amidst the conveniences of technology has caused us to suffer a spiritual death, to feel alienated, empty, without purpose and direction. And that may very well be the case. But nevertheless a radical distinction must be made: the need for transcendence experienced by most human beings prior to modern times was a very different one from that which is claimed to exist today. It is not likely that the human race before our time, despite its life dominated by religions and churches and yearnings for transcendence, was a jot more spiritualized than it is today. For if the connection between the growth of industry and the decline of religion is a real one, the earlier spiritual longings appear as an escape from humanity's vulnerable position in its battle with Nature. It was not that humankind's esthetic sensitivities to the idea of the good and the idea of the beautiful were any more developed in past history; rather, its need to escape from an intolerable physical life was infinitely greater than ours, for our physical lives are not very oppressive. That "other," "better" world offered by religion could not have been *worse* than the "real" one, even in the duties that it required on earth, and as a mere fantasy it offered extreme gratification. When I speak of humanity's previous need for transcendence over the insupportable conditions of physical life, I do not refer to the needs of great creative people—artists, thinkers, craftsmen—who by their very temperaments can never be satisfied with any status quo. I speak of the masses of people whose spiritual lives were necessary to make their physical lives endurable and who, had choice been possible, would certainly have preferred physical comforts over spirituality. This situation does not for the most part now exist: television and toilets have made the need for God supererogatory. Western humanity does not generally live in fear of Nature, except when earthquakes or cancer strike, for it is mostly unaware of a connection with Nature that has been artfully concealed by modern technology. Almost every deprivation has its accessible remedy, whether hunger, cold, illness, or mere distance; and there is rarely a need, except at a few moments during one's lifetime, to go crying either to Papa or to God the Father.

If a need for transcendence does exist today, a question that I am not here pursuing, it is in any case not the same need that formerly was so widespread. It is a need based on satiety and not on deprivation, and it does not seek a haven in another world but rather a more beautiful version of this one. What I *am* concerned to examine here is what has happened as a result of the Industrial Revolution to humanity's conception of its relationship with Nature and what has become the present form of the old mind-body duality.

To the average child of the United States in the present day, Nature is indeed a great mystery, not insofar as it is incomprehensible but insofar as it is virtually nonexistent to his perceptions. Not only do most children obtain without delay the nurturing commodities for a satisfied bodily life, but they are rarely in a position to experience a connection between the commodity that fills their need and its natural source. "Meat" consists of red geometrical shapes obtained in plastic packages at the supermarket, whose relationship to animals is obscure if not wholly invisible. Houses are heated by moving a thermostat, and clothes are washed by putting them into a washing machine. Even the child's most primitive natural functions are minimally in evidence, and it is not surprising that various psychological problems turn up later on in life when our sensual nature has in some way been concealed at every point by technology. (I recall a student who once remarked that she had no desire to venture out into the country to "enjoy Nature" when she could see all the trees she wanted on TV.)

The reader should be assured that I am not engaged in presenting these observations in an effort to make the familiar attack on "technology." I have no personal objections to meat in plastic containers or flush toilets and air-conditioning. In fact, I like them very much. I have no desire to hunt animals, to chop down trees for firewood, to use an outhouse, or to have smallpox. I have no interest in a "return to Nature," which strikes me as an especially decadent form of estheticism, like an adult of forty pretending to have the innocence of a child. My consciousness as a person living at a particular stage of history cannot be wiped away by a decision to perform a Marie Antoinette. I would much prefer to listen to music or work in the garden than to struggle for survival. I have presented a picture of a hypothetical child who sees no relation between the red glob in the plastic carton and the animal from which it came, not to attack either technology or modern techniques of child raising. What I am trying to do is to present a picture of humanity's current relation to Nature.

With Nature barely in evidence and our physical needs satisfied beyond what could have been imagined one hundred years ago, the human mind would appear to have arrived at a state of altogether new autonomy and independence—not this time the independence of a mind that has given up all hope of dominating Nature and satisfying the flesh and therefore seeks in desperation a haven in Abraham's bosom; rather, this time, a mind so assured of its domination of Nature and its capacity to satisfy the flesh that it seems to be borne up on its own engine of will, cut off from any nurturing roots in the earth. Mind, now soaring not on wings of fear but on sturdy pinions of volition, can say to Nature, "Get thee behind me, Satan!" Do not presume, it would say, to interfere with my self-determination, for if you do, I will flip on the air-conditioning, switch on the electronic air cleaner, swallow down the antibiotics, spread on the weed killer, inject the flu vaccine, fill up the gas tank.

But while all of this newfound mental assurance has been building up, when *Homo sapiens* has finally found a home in the world, when he feels he is lord of all he surveys, when he no longer needs to have his spirit stroked by the right hand of God—a new "trouble" (which I put in quotation marks because it is thought by some to be purely imaginary) rears its ugly head: humanity's nurturing environment threatens to stop nurturing and to start killing.

One opens the newspaper each day to find four or five articles whose burden is that pesticides contaminate the food of farm animals in Michigan; Kepone is being dumped in waterways, asbestos fibers in Lake Superior; poison gases render uninhabitable a village in Italy; the Parthenon is decaying faster in ten years than in the previous thousand because of automobile exhausts; ozone and sulfur dioxide increase mortality rates in Chicago and Los Angeles.

Although we had been taught in our high-school science classes for decades that neither matter nor energy could be created or destroyed, suddenly it dawns upon someone that the refuse being dumped into the oceans and atmosphere for years and years in ever-increasing quantities does not "go away." Where was it supposed to go? Suddenly, the human race has been put into the position of affluent teenagers who dump beer cans from their moving sports car and then drive off. The cans appear to have vanished, but no, there they are, astoundingly enough, rolling around the neighborhood where they had been dumped. And when the teenagers arrive home, they find other beer cans dumped by other teenagers. The neighborhood is a place of beer cans; the ocean a place of toxic effluents; the sky is vaporized garbage. And to add insult to injury, man's unconquerable mind turns out to have a mouth, through which it is fed; and worse still, it is being fed garbage. Its own!

Before continuing, let us stop for a moment to see where we have been: in the early days, humankind had no power over Nature and turned, instead, to its mind and its gods for consolation. Meanwhile, this mind produces a technology that enables its body to be as strong as the gods, rendering the gods superfluous and putting Nature in a cage. Then it appears that there is no Nature and that humanity has produced virtually everything out of sheer ingenuity, and it can all be bought in a supermarket or a discount store, wrapped in plastic. By now, humanity is scarcely aware that it is eating animals and producing wastes or that the animals come from somewhere and the wastes are headed somewhere. This "somewhere" turns out to be, practically speaking, a finite world whose basic components cannot be created or destroyed although (and here is the shocker) they can be turned into forms that are unusable by people. As more and more of these basic materials are rendered unusable, it becomes apparent that humankind has failed to see that now, as in the past, the roots of its being are in the earth; and it has failed to see this because Nature, whose effects on people were formerly *immediate,* is now *mediated* by technology so that it appears that technology and not Nature is actually responsible for everything. This has given people a sense that they mentally and voluntarily determine the ground of their own existence and that the body is almost a dispensable adjunct of their being. This is modern humanity's own peculiar mythology: the myth of voluntary omnipotence. It is the contemporary form of the Faust legend, a legend which in all of its variants ends the same way. Nowhere is this modern version of the Faust myth so apparent as in the words of industrial corporations who attack the basic conception of environmental protection. If the classic flaw of the tragic hero is overweening pride and a refusal to acknowledge his own finitude, the contemporary Faustian attitude is archetypically struck in the advertisements of steel and oil companies protesting that "stagnation is the worst form of pollution." The current terminology of doublespeak can be seen in the modish word "trade-offs," a concept which would admirably serve as the basis for present-day tragic drama. One would suppose from such talk that modern industrial corporations, with their fears of economic stagnation and their estimate of clean air as an unaffordable economic luxury, were Shelleyan Prometheuses, defending humanity's sublime aspirations in the face of a tyrannical and boorish Zeus.

The continual appearance of the concept of "trade-offs," in which one sacrifices the "luxury" of an uncontaminated environment in order to permit economic "progress," brings to mind a cartoon that I saw years ago, before anybody ever heard of the environment: two emaciated and threadbare prisoners are bound with manacles and pedicles to the middle of a wall about four stories high in an immense featureless white room. Flailing upon the wall, about two stories above the ground, one enfeebled prisoner says to the other, "Now here's my plan . . . ." Is this not an emblem of modern man? Oblivious of his roots in the earth or unwilling to acknowledge them, intent only upon the desires of his unconquerable mind, he refuses to see that his well-nurtured body and Faustian will are connected by fine tubes—a "life-support system," if you wish—to the earth. Can those Faustian thoughts continue without a narrowly prescribed nutriment for the body, a nutriment prescribed not by that Faustian mind itself but by a biological determination that has been *given* rather than *chosen*? Are not the limitations once described as the will of God and as "grace" as much limitations now as they have ever been in the past? Unless humanity can create itself, unless it can determine its own existential nature, how can humans talk—absurdly, madly, derangedly— about "trade-offs" with the environment or "negotiations" with Nature?

Can one negotiate with the *données* of human existence? Even a Promethean Sisyphus needs food to push his rock.

I recently had occasion to publish two essays describing the traumatic effects that polluted air has had upon my wife and me during the past six years, one of my major points being that we are not "cardiac and respiratory patients" but normally healthy people whose lives have been radically altered by industrial emissions since we came to live in the Chicago area. One of these essays, a brief account of our experiences that appeared in the *New York Times* and was subsequently reprinted in other newspapers, brought me a number of interesting and varied responses from readers. A letter that particularly struck me read as follows:

> Dear Sir:
>
> Since all of the environmentalists who worry about pollution are also consumers of the products of these belching plants (the automobile for instance by which you reach your farm), what is the answer? Do we cut off our noses to spite our faces? Do we destroy our economy: eliminate many necessities of life; go back to living in tents for the sake of clean air? The answers are complex.

This was a profoundly disturbing letter. The writer was by no means insensitive to the problems of our time; she saw that a complex dilemma is involved; and she was obviously very concerned about the entire affair. Yet her expression "for the sake of clean air" is a familiar one and reveals that the heart of the problem has not been grasped. For when she asks, "Do we eliminate many of the necessities of life for the sake of clean air?" one wants to know: what are the necessities of life in comparison with which clean air cannot be regarded as a necessity? But to ask this is to raise a purely rhetorical question, for the problem is really an ontological and not an ecological one.

When the writer refers to the "necessities of life" one must ask what it is that she means by *life,* and I am proposing that by "life" she means her desires and her will; by the "economy" and "necessities" she means those things which support her mind's conception of itself. There is not a body in sight. She sees steps taken to preserve the environment as actions "for the sake of" clean air. She does not see them as "for the sake of" her own biological existence. *Somehow,* she is alive: she eats food, drinks water, breathes air, but she does not see these actions as *grounds of life*; rather, they are acts that *coincide* with her life, her life being her thoughts and wishes. The purity of the elements that make her life possible is not seen as a condition of existence. Instead, the economy, the "necessities," and not "living in tents" are what matter. *That* is life. Her existence on earth somehow takes care of itself, so why sacrifice the "necessities" of life "for the sake of" the superfluities, like "clean air"?

The pattern of thought this letter reflects becomes clearer if we make some substitutions: "Do we eliminate necessities of life for the sake of clean air?" could equally well be presented as "Do we give up smoking for the sake of avoiding lung cancer?" since smoking occupies the role (for those who feel they must smoke) of a necessity of life and "avoiding lung cancer" occupies the position of "for the sake of clean air." However, "avoiding lung cancer" can be more clearly stated as "remaining alive," which would then yield the question: "Do we give up smoking for the sake of remaining alive?" And in a final transformation we may obtain: "Do we give up the necessities of life for the sake of remaining alive?" I offer that as the paradigmatic question behind all of the similar ones that people ask. On the surface, we are faced with a paradox: how can someone ask whether it is necessary to give up a condition of life in order to remain alive? But the paradox evaporates when we realize that the "necessity" is no necessity at all, from the viewpoint of our biological existence. Rather, the "necessity" (smoking, the present economy, etc.) is a mental stance, a wish, that in fact is inimical to the survival of the body that would make it possible to continue to fulfill the wish.

We are able to see that this is a variant of the traditional mind-body problem, the view here being that man is his mind, that man is his thoughts and wishes. But man's sublime mind (not to mention the very unsublime wishes described above), while it may wander at will through the universe and be connected to the heavens at one end, is connected at the other to the earth. As free as that mind may appear in its wanderings, thoughts rely on calories, because they are fueled by the same metabolic processes that make all other human activities possible. A thought may have no weight and take up no space, but it exists as part of a stream of consciousness that is made possible by food, air, and water. Every moment of a person's existence as a human being is dependent upon a continuous burning up of energy, the classical tragic conflict consisting of a mind that is capable of envisioning modes of existence that are not supportable by a human engine thusly fueled. The confidence of Oedipus that he could out-wit causation provides the model for the present environmental dilemma. But there is little that is new about this dilemma besides its peculiarly contemporary terms. The struggle between the "necessities of modern life" and the "environment" is the age-old struggle between the individual will and the universe, the substance, in other words, of classical tragedy.

Thus "the problem of the environment," which many people persist in viewing as a peripheral arabesque drawn around the "important" concerns of human life, must ultimately be seen as a central philosophic and ontological question about the self-definition of contemporary humanity. For all one's admiration of man's unconquerable mind and its Faustian aspirations, that mind would seem to be eminently conquerable, particularly by itself. It is, after all, a very frail vessel, floating upon a bloodstream that is easily contaminated by every passing impurity: alcohol, nicotine, sulfur dioxide, ozone, Kepone, DDT, sodium nitrite, red dye #2—the list appears endless. As much as at any time in the past, however, the human relationship with Nature is nonnegotiable. Perhaps within a certain narrow range *Homo sapiens'* constitution is susceptible to adaptation, but in the light of the innumerable and arbitrary concurrences that make human life possible, our adaptability seems very limited indeed. In the past, humanity's Faustian aspirations were seen against the background of its terrifying weakness in the face of Nature. Today, humankind's Faustian posturings take place against a background of arrogant, shocking, and suicidal disregard of its roots in the earth.

CHAPTER FOUR

Air and Being

The Psychedelics of Pollution

Surely, nothing is more reproachful to a being endowed with reason, than to resign its powers to the influence of the air, and live
in dependence on the weather and the wind for the only blessings
which nature has put into our power, tranquility and benevolence.
This distinction of seasons is produced only by imagination operating on luxury. To temperance, every day is bright; and every hour is
propitious to diligence. He that shall resolutely excite his faculties,
or exert his virtues, will soon make himself superiour to the seasons; and may set at defiance the morning mist and the evening
damp, the blasts of the east, and the clouds of the south.

SAMUEL JOHNSON

He had, till very near his death, a contempt for the notion that the
weather affects the human frame . . . Alas! it is too certain, that
where the frame has delicate fibres, and there is a fine sensibility,
such influences of the air are irresistible. He might as well have bid
defiance to the ague, the palsy, and all other bodily disorders. Such
boasting of the mind is false elevation.

JAMES BOSWELL

A LAS INDEED! The influences of the air are even more irresistible than
Boswell's prescience could have envisioned. After Darwin, Marx,
and Freud, the arena of human freedom has come to seem painfully
shrunken. And after contemporary environmental studies, even less remains. But recognition of environmental constraints upon our behavior
can at least inform our options, as we come to see how many "choices"
are actually made for us by the nature of things.

My own knowledge of these matters springs from the personal experience of having lived in the dramatically polluted environment of northwest Indiana, not far from the steel mills and power plants that line the shores of Lake Michigan from Chicago to Michigan City. And since the essence of that knowledge introduces what might be called an existential environmentalism, its accumulation from firsthand experience lies at the heart of the whole affair. For despite the almost unbelievable development of knowledge about environmental issues during the past two decades, the *personal* realities of the problem of bad air remain almost unexplored. The dependence on technological means of measuring air quality has presented a very skewed picture of what it means—what it feels like—to be a person amidst pollution, and this picture continues to leave the erroneous impression that pollution is a somewhat abstract problem that affects other people rather than me myself or that at bottom it is mainly an esthetic nuisance. Considering the limited extent to which most people are aware of how pollution concretely affects them at this very moment, public support of environmentalism exhibits a not-so-common instance of disinterested human concern. We have repeatedly been told that polluted air is bad—for plants, for animals, and for other people—and we have responded accordingly. But how, we need now to ask, is pollution bad *for me*—as I take this very breath, as I try to go about my daily tasks with the sense, probably illusory, that I am in control, now, as I am reading these words, not merely in some nebulous future when I may learn that I have become a casualty?

The media, spurred on by government agencies, have helped to foster this unbalanced picture of pollution as bad for others. The constant warning is that cardiac and respiratory patients should stay indoors today (in their private air supplies?), or possibly that joggers along Lake Michigan should take it easy because of the dangers of Ozone Alley, the lakefront from Waukegan to south Chicago. But presumably, the rest of us are fine. Television weather reports rarely mention pollution, except in places like Los Angeles, even though there is literally no such thing as weather separable from the vast transcontinental air currents that conduct emissions hundreds, even thousands, of miles. And when they do mention it, they mislead us by reporting that air quality in Chicago was fine today, while ignoring the fact that pollutants had blown out into the suburbs or countryside, to whose residents it was far from fine, even though they have been told otherwise. The so-called urban and suburban "haze," which sounds so harmless and romantic, the snow squalls that plague southwestern Michigan and northwestern Indiana even though it is snowing nowhere else, the stench of sulfuric acid emanating from suburban snow banks—these are not simply "weather." For Commonwealth Edison and U.S. Steel produce as much of it as God.

To further confuse our understanding, the data upon which reports are made about health problems are gathered mainly from hospital records, that is, from the most extreme and dramatic cases. This helps to explain why there is so much misleading emphasis placed upon cardiac and respiratory patients: these are the cases that end up in hospitals and thus are the most visible. But the effects of air upon the general population are rarely discussed—and rarely will be unless reports are made about people who are not in hospitals. Given the nature of technological society, it is very difficult to obtain information about matters that cannot be easily fed into a computer. But ordinary people show every sign of being affected by pollutants as much as "sick" people, even if they don't end up in a hospital. And these maladies occur not just among dwellers in smoky cities, but among people who live in the suburbs, in the country—in a word: everywhere.

In fact, there are few if any "safe" places that are so far from pollution sources as to be exempt from today's intimations of mortality. When my wife and I moved from our rural farm fifteen miles south of Gary, Indiana, to our present suburban home about sixty miles northwest of Gary, we naively believed we would be far enough from the steel mills to escape the air that was making us ill more than half the days of the year. But Chicago's northwest suburbs, superior as they may be to northwest Indiana, are polluted enough. Southeast winds blow Gary's emissions deep into Wisconsin, while Waukegan's power and steel plants send their devastating pollutants to the already overburdened Illinois atmosphere whenever winds are from the north and east. With Joliet sending its own contributions on southwest winds, very few days of clean air are available in the greater Chicago area. And they may be getting fewer as coal becomes the latest panacea.

Up in Madison, Wisconsin, 160 miles or so north of Indiana, one can still discover the plume of bad air from Gary-Chicago if the winds are right, with Milwaukee sending off its own toxic clouds. And if that seems far, on one of our visits to the English West Country, the Cornwall of legend and song, my wife and I began to experience the familiar physical and mental symptoms of bad air, with much disbelief and despair, only to learn a few days later from the London *Times* that the Ruhr Valley's heavy industry in faraway Germany is a major source of British air pollution. In the United States, the vines of grape growers in western New York State suffer marked damage because of emissions from Gary; New York City is said to receive much of its bad air from Pittsburgh, Gary, and Birmingham; while New England inherits it all as acid rain. Indeed, the *New York Times* of February 8, 1979, reported: "Smog and dust from industrial Europe and China may account for a mysterious haze that hangs over Alaska, Greenland, and the Arctic Ocean every spring, according to analysis of atmospheric particles that are assumed to cause haze." Whither then escape? Denver has had its day, Los Angeles and Riverside are cesspools of pollution, Ohio a nightmare, Phoenix in decline, and Waukegan bad for your health. The circles increase in contemporary hell.

There are a number of ways of recognizing air pollution without specialized instruments. Visually, the most dramatic signs are familiar by now to most people in city or country: orange-brown smog that obscures almost everything within range and which, in its lesser presence, is euphemistically referred to as "haze." Apart from the more dramatic cases—inability to see the Hollywood Hills, amorphous quality of skyscrapers just across a New York street, even in sunlight—smogs and hazes are not so easy to notice unless one lives in the country or atop a high-rise. Distance is required to get an accurate picture, since the sky directly overhead almost always seems to be clear, giving the comforting impression that one's self is almost never amidst the pollution. As one drives into the city the nimbus of smog seems far ahead, hovering over tall buildings, appearing to recede on arrival, until—miraculously—the city does not seem polluted when you are in it! But if there is haze on three sides one must obviously be in it, however clear overhead. Furthermore, some of the most intensely polluted days consist of gaseous emissions with little particulate matter, and only a careful scan of the horizons hints at the presence of sulfur dioxide or ozone in what otherwise appears to be a clear and sunny sky.

Smells, another sign of emissions, are harder to detect because one's sense of smell is dulled very quickly. Often, a sudden exit from the house will reveal stronger concentrations of odors that had infiltrated too gradually to be discernible inside. Odors suggestive of asphalt or tar as well as oil and "baked potatoes" are common byproducts of steel mills and petroleum plants, and the smell of ozone on summer mornings can be detected in and around urban areas with heavy traffic from automobiles.

But it is the physical and psychological signs of air pollution that are the most important and for which the visual and odorous merely provide confirming evidence of the extent to which one's total being is shaped each day by the particular chemical mix of the air. The range of these symptoms is great, and most often they are found in combinations rather than singly. Though many may also be found in connection with maladies unrelated to the air, it is the circumstances of their combination that enable them to be traced to industrial and automobile emissions. When evidence of sight and smell accompanies these experiences, when the wind is blowing from the direction of major pollution sources, when oneself and friends feel these maladies at the same time, and when the symptoms vanish with a shift in the winds, only sheer perversity can cause one to fail to take the hints.

Although coughing and burning eyes are most familiar, and although official warnings usually stress respiratory ailments, these discomforts are only the most obvious and operatic of the effects of bad air. More widespread and insidious than these are a broad variety of headaches, often combined with nausea and dizziness, especially on arising in the morning. One is likely to have slept through the night as if under sedation, awaking dizzy, drugged, and in a stupor. This commonplace "inability to get up in the morning" is not just a *donnée* of human life, however, but a gift of industrial society. Once up, one may feel unsteady, heavy-headed, with a growling stomach and heartburn, preoccupied during breakfast and inattentive later in the day, unable to focus or maintain a clear train of thought. Nor does it matter if one has had six or sixteen hours of sleep, since the problem is not lack of sleep. If one is suffering from hunger and heartburn, the hunger does not go away, even after a decent breakfast, but rather insistently gnaws, producing an insatiable craving for high-carbohydrate junk foods. On badly polluted days people wonder why they are eating all those potato chips and candy bars when they have had their usual breakfasts, and the junk food machines will be whirring away. Thoughts of dinner start crowding the mind early in the afternoon. In the highest realms of diplomacy, the august diplomats, bowed by the weight of international affairs, can only think of sweet sherry and scones. Reading errors, typing errors, carpentry mismeasurements start to increase, along with an ill temper that seems to be caused by nothing in particular. The more sedentary the activity, the more one is at the mercy of out-of-focus intellect, though joggers are warned to take it easy.

Like the extreme difficulty getting out of bed in the morning and the curious hunger even after eating, the pseudo-cold is not yet generally associated with badly polluted air, but only a little attention is needed to make the connection: a sore throat suddenly seems to be developing, one's arms and legs, joints and muscles begin to ache and seem sore and fatigued, and one's head feels heavy as in the onset of a cold. One wishes simply to fall into bed and give way to the miseries of a cold. But then next day, magically, all of this disappears. One forgets that a cold was even settling in. The winds have shifted. Where are the colds of yesterday? They have literally blown away. For me the most striking example of this took place a few years ago in London. I had arranged with a friend early in the week to go on a weekend walk in the country, but by midweek I began to feel completely overtaken by cold symptoms and lethargy. When I took to my bed, I phoned to cancel the outing and learned that both my friend and her office coworker had begun to feel exactly as I did on the same day and felt that way still. During the following week we reported to each other that our "colds" had gone away on the same day, when indeed the persisting winds had shifted direction.

Depending on the industrial mix, one is apt to have nausea and dizziness, pains in the hands and feet, chest pains, heartburn and gas, muscle aches, burning eyes, lethargy, and headaches, to mention common signs. But beyond the physical are depression and dispiritedness about nothing in particular, short temper, irritability, aimlessness, a tendency to quarrel, an inability to read, a general inattentiveness, and a despair of the possibility of human happiness. One is conditioned to a frightening degree by the day's particular—and particulate—mix. A midday shift in the wind can change one's philosophy as dramatically as one's cold symptoms.

Reflecting on all of this, one is led to ask why, after all, should the experience of "just feeling out of sorts today" be exempt from the causality that lies behind all other kinds of phenomena? These experiences must have their causes like any other, however hard to pinpoint. And so, when it finally becomes apparent that transitory mental and physical states are not just causeless random happenings, the discovery of pretty consistent patterns becomes not only practical but inevitable. Because periods of malaise can range from only a few hours to many days in a row, depending on atmospheric conditions, it is rarely possible to have these experiences diagnosed by a physician. The ailment is gone by the time the appointment day has come around; yet even if it were still present, it is highly unlikely that the average medical doctor would provide a correct diagnosis. For most physicians know little or nothing about the psychological or physiological effects of pollution. To further complicate things, the sudden recoveries from environmentally induced ailments cause the sufferer to forget about them as soon as they have gone away. When they recur, the same kind of cycle is apt to take place. One never gets any closer to the solution of these problems.

So one must try to solve them by oneself. Learning to ascertain the correlation between physical/psychological symptoms and air pollution requires an awareness of the principal sources of pollutants for any given section of the country. A representative picture can be derived from the Chicago area: the entire south end of Lake Michigan produces massive quantities of air pollution, much of which is often strikingly visible and smellable while driving on the Chicago area expressways. With southeast winds, to use one limited illustration, a vast cloud of pollutants starts out from a fairly narrow area south of the city (from a triangle that includes East Chicago, Gary, and Michigan City) and spreads in an ever-widening wedge over Chicago and its northwestern suburbs, a wedge that is easy to see while driving on the tollways that circle the city. If the wedge remains narrow, one can drive in and out of it, sometimes more than once during a trip in and around the city. When it is broad, this wedge spreads out well into Wisconsin. But if you happen to be close to its starting point, where it is very narrow, it is actually possible to miss its effects altogether even though literally millions of people are living in its shadow at that same moment farther away.

A drive around the perimeter of the city on the Tri-State Tollway can be an educating experience, for it is possible to move between sunny, clear, and beautiful skies at one end (i.e., the Indiana or the Wisconsin border) and dark, foul smog, or even a very confined snowstorm at the other, all within an hour and a half. If one is alert, it is possible also to observe the physical and mental transformations that may take place as one enters and leaves the different mixes of air. Headaches and dizziness can appear and disappear as one rounds the large curve from O'Hare airport to Hammond, Indiana, accompanied by the "baked-potato" smell, the Sherwin-Williams paint factory smell, the oil refineries smell, and the smell of *real* potatoes from Jay's potato chip factory.

Although the effects of being downwind of a pollution source can be very pronounced, it is only when people compare notes about how they feel that illuminating causal connections can be made. When out-of-town friends were visiting me on what seemed to be a beautiful, cool, clear summer's day with light northeasterly winds from Waukegan (a dependably bad source of pollutants), they suddenly announced that they both felt so miserable that they would have to nap for a while. It must be air pollution, they told us, because they had headaches, felt drowsy and lethargic, and could hardly keep their eyes open. And to top it off, they were in very low spirits, verging on despondency. We had talked about such things with them before, and now had to agree that the air was pretty bad. My wife and I had struggled out of bed that morning and tried to be lively hosts even though we felt lifeless, dazed, and unfocused all during the day. Misery had the company it so often wants, which helped to cheer us all up.

A new dimension can be added to the old philosophic chestnut about free will: it is not merely one's genes, one's prior psychic history, one's parents, social class, etc., that determine one's accomplishments, moods, and perspectives. It's also the chemical mix of the very air one is breathing at any given moment, for breathing such air is a counterpart of eating food contaminated with pesticides or drinking water laced with asbestos fibers or polychlorinated biphenyls (PCBs), except that the effects are often very immediate. How "free" is a creature whose worldview at a given moment has literally been concocted miles away in the vat of a steel mill? If one can be drugged without pills, soused without Scotch, depressed without precipitating psychological events, irritable without irritants, and pessimistic without philosophy; if one can be hungry without fasting, exhausted without having expended any energy, and afflicted with heartburn and indigestion without recent food, then what does it mean to have a mind or a will of one's own?

Seen in this light, various kinds of experience shed their metaphysical mists and encourage a sordid behaviorist perspective. For instance, in the

Chicago Tribune of January 20, 1980, the food writer Carol Haddix made some routine observations about her uncontrollable hunger:

> The hunger pangs were beginning. It usually happens around 10 a.m. whether breakfast is eaten or not. It becomes difficult to work with that nagging stomach rumbling away. Coffee or a cup of tea works to stop those pangs for a short time, but your body knows better. "Where's the food?" it cries.
>
> I usually last until about 10:30 or 11 a.m. before I make a mad dash to the junk-food machine and wolf down a disgusting candy bar or a bag of potato chips. Oh, if my mother only knew.

If only the writer herself had been in a position to know: a polluted morning in Chicago, an insatiable craving for high-carbohydrate foods despite breakfast. A check of the day's pollution readings would doubtless reveal more than one's seeming-wise body that supposedly "knows better." Though perhaps the body does know better, since air quality is causing it to crave the sort of nutrients that provide quick energy.

When junk food machines are cranking out their chocolates, when professors wonder why their classes are afflicted with lassitude and inattentiveness on Monday, while they are very lively on Tuesday despite less interesting subject matter, when office workers can barely do their typing—when all this is finally observed and tallied up, new knowledge has become possible.

As a coda to these reflections, on its science page of October 6, 1981, the *New York Times* reported some remarkably interesting fruits of research done by Jonathan M. Charry of Rockefeller University and Frank B. W. Hawkinshire V of New York University. In an article that appeared in the *Journal of Personality and Social Psychology* (41, no. 1 [1981]), they discuss the harmful effects upon personality and behavior produced by an excess of positive ions in the air. "In environments where the ion balance is constantly shifting because of industrial activity, auto emissions, and high-voltage lines, understanding the basis of susceptibility to adverse effects of ions is likely to have considerable importance." They go on to remark that "atmospheric charge is a fundamental part of the air we breathe. Since altered ion concentrations can result not only from changes in weather conditions but also from the presence of pollutants, high-voltage lines, and radioactivity. . . it is becoming evident that many elements in the physical environment can be highly irritating or stressful, leading to changes in social behavior." The *Times* piece emphasizes the effect of these positive ions on "people's mood and health, even precipitating suicides, crimes and accidents," as well as such ordinary reactions as increased tension and irritability.

Boswell's intuitive awareness of more than two hundred years ago may now be ready for general circulation.

CHAPTER FIVE

Ecocriticism's Genesis

A LTHOUGH WILLIAM RUECKERT is usually cited as the only true beget-ter of the term "ecocriticism" in his 1978 essay "Literature and Ecology: An Experiment in Ecocriticism,"[1] like just about everyone else I did not come across the term until much later. For me personally the only true begetter was Cheryll Glotfelty (formerly Burgess), from whom I received—retrospectively considered—a stunning letter in May of 1989, a form letter in fact that had also been sent out to two hundred other authors. It began,

> . . . I'm a PhD candidate in American literature at Cornell University, purportedly finishing up my dissertation on representations of nature in the American women's literary tradition. The question that fires me incessantly is this: how can one, as a literary critic and teacher, contribute to the ecological health of the planet? It seems to me that ecological concerns are so pressing that they ought to eclipse every other concern. If I can't find a way to approach literature ecologically, then I will have to abandon this profession as frivolous.

Cheryll realized literary-environmental writings were scattered far and wide, as were the practitioners of such writing, most of whom probably felt like solo voices crying in the wilderness. To support this claim, Cheryll attached an amazing bibliography that she had compiled of what would come to be called "ecocritical" writings—it was to the authors of these that she mailed her letter as a twofold plea for help: to add to the bibliography and to help her to produce an anthology of the best of it. Her goal after accomplishing this, she concluded, was "to be the first professor of Literature and the Environment" once she finished her degree at Cornell and entered the job market (which in those days had not yet become bleak for the humanities). At the bottom of her form letter she appended a handwritten note, alluding to her enthusiasm for "From Transcendence to Obsolescence" and psyching me up for the sequels that were to follow.

I marveled that Cheryll had discovered all three of the essays that pre-
cede this chapter, since in those days there was no category in which to fit
them and, as far as I knew, they had not been picked up in any bibliography or citation index. When I asked about this later on she reported that
a professor friend of hers was using as text the *Norton Reader,* which had
anthologized the "Transcendence" essay, appropriately positioned (in keeping with my primordial neuronal mix) between Plato and Woody Allen.
Somehow, after reading it, she pressed on to find the other two.

When I responded to her letter with enthusiasm, I received powerful
encouragement. After extremely high praise of my essay, she added a
postscript in which she reproduced for me the note she had written to herself after reading it: "Although acutely aware of ecological balance, this
essay is inveterately anthropocentric still, showing little regard or concern
for how human activities affect other species, focusing instead on how
human activities, given biological realities, backfire to affect *humans.* I
think even Fromm's worldview stands to be enlarged."

How could I have resisted such a consciousness-raising challenge?

The letters that followed sprung more surprises, such as the invitation
to be coeditor of the purported anthology. I agreed to the editorial job if
I could give her most of the credit for the book (which I did), and in the
winter of 1990 I came up with the idea of proposing the very first ecocritical session to be held at the Modern Language Association convention.
With a call for papers in the *MLA Newsletter* for a session on the greening of literary studies, we were launched into the unknown. The response
was encouraging: many more potential papers were offered than could be
accommodated. I chose the most promising, wrote a formal proposal—
and MLA turned it down. Some of us sent angry complaints to the MLA
office, asked to see the judges' reports, and were incensed even further by
the reports' flimsiness. (This happened once again when I more recently
proposed a session on Darwinian literary theory—and the excuses were
even flimsier.) Undaunted, we tried again the following year with an almost identical proposal and the very same panelists, this time with mysterious success. Not knowing what to expect, I asked for a meeting room
to accommodate an audience of twenty-five.

On the 29th of December, however, I was astonished to see that our
meeting room at the San Francisco Marriott was filling up rapidly and
would probably be too small to accommodate the rapid influx of attend-

ees. Managing to negotiate a room switch even as the crowd was pressing in, we ended up with more than a hundred enthusiastic auditors. Cheryll, as first speaker, introduced the subject of ecocriticism—and the movement already under way began to take on the look of an institutionally sanctioned discipline. Not wanting to lose these mostly newfound allies, we passed around a sign-up sheet so that we could keep in touch.

Not long afterward, the Western Literature Association, which had been in existence since 1966 with particular literary emphasis on what had come to be called "nature writing," held their October 1992 conference in Reno. A special meeting took place to plant the seeds for a new organization to be devoted to nature writing. Michael Branch and Scott Slovic, editors of *The ISLE Reader: Ecocriticism 1993–2003,* tell the story. "All agreed that the time was right to . . . promote environmentally oriented work in the humanities . . . Although the journal *ISLE* [Interdisciplinary Studies in Literature and Environment] had yet to produce its first issue, Cheryll proposed that we adopt a version of the journal's name and refer to the new community of scholars as the 'Association for the Study of Literature and Environment.' We would abbreviate the name as 'ASLE' (pronounced 'AZ-lee')."[2]

By the next year, ASLE was a functioning organization, and Cheryll and I began to correspond more animatedly about the contents of what came to be called *The Ecocriticism Reader: Landmarks in Literary Ecology.* Two years earlier, Cheryll had already remarked in a letter to me that "despite the fact that you and I are now thoroughly drenched in ecocritical thought, the idea of combining literature and ecology is still unheard of and confusing to most people . . . One of the most significant things about our new book is that it *will* be inaugural, and, as such, will help to establish and define a new field of study. It will be *important.*" Cheryll was indeed right about the perceived strangeness of the nature of ecocriticism. When we finally got our act together enough to produce a manuscript, one of the first turndowns from publishers was from Cambridge University Press in 1993. "From the evidence I have seen," the editor wrote, "the status of ecocriticism as a scholarly discipline is still under negotiation." Conceding the growth of environmental interest in the academy, the editor nonetheless continued, "The suppleness and relevance of the issue, which have allowed it to apply in critiques ranging from the economic to the literary to the theological, have yet to produce a coherent praxis for 'literary ecology.' This casts a dubious light on *Ecocriticism*'s readership, and by extension on the viability of the project as a publishing venture for us." The University of Georgia Press, given its predisposition to publishing environmental studies, turned out to be much more adventurous and prescient. They snapped up our book very quickly and came up with the title by which it is known. (In 2005, the press reported to me that after ten years in print our book was still a bestseller.) Besides collecting eco-writings that had been widely scattered, unavailable, or unknown, we encouraged several young academics who performed at our MLA session to expand their talks into essays, which we then included in the collection (and these have been among the most cited). Although Cheryll and I never envisioned our anthology as a textbook, it became the founding and canonical work, well-reviewed, frequently chosen for use in courses, and cited in almost every article and book that deals with ecology and literature. In 2005, Georgia asked if we would be willing to produce a new edition of the reader with updated selections, but we declined. Since its publication in 1996 there had been a deluge of anthologies and monographs that made a second edition superfluous. Ecocriticism is today a bona fide field of study with unofficial headquarters at the University of Nevada, Reno, where Cheryll, Michael Branch, and Scott Slovic have pioneered a program. And in 1990, Cheryll indeed became the first official Assistant Professor of Literature and the Environment.

For me, learning from Cheryll's introductory letter of 1989 that I was an ecocritic was an event much like the famous case of Molière's Mon. Jourdain, who was astonished to discover he had been speaking prose all his life. But of my first three ventures into ecological writing (the three preceding chapters) the only one of the group that can be considered ecocriticism is "From Transcendence to Obsolescence," which treats ecological themes as variants of classical literary preoccupations. "On Being Polluted" was essentially *ecology* from a humanist's sensibility, but insofar as its hidden real theme was the effect of air pollution on the body and thence on consciousness, it marked the unwitting germination of a consciousness-oriented subtext to everything ecological that followed in my thinking. "Air and Being" was a much more explicit account of the environment/body/mind interrelation that became increasingly dominant each time I returned to seemingly "environmental" issues. Thus, most of what follows in this book shows ecology morphing bit by bit via Darwinian themes into consciousness studies. "After Darwin, Marx, and Freud," I had written in 1983, "the arena of human freedom has come to seem painfully shrunken. And after contemporary environmental studies, even less remains. But recognition of environmental constraints upon our behavior can at least inform our options, as we come to see how many 'choices' are actually made for us by the nature of things." (Compare that rather bland statement with the final chapter of this book.)

Writing about the environment hardly constituted a new genre. One could trace it back at least as far as Theocritus, through the Romans, the Medievals, the European Enlightenment, the Romantics, the American Transcendentalists, and so forth, not to mention the literature of the East. But the literary mode is almost always that of "nature writing," as it came to be called later on. When the activities now subsumed by ecocriticism generated ASLE, *ISLE,* and the various meetings held by the Western Literature and Modern Language Associations, the emphasis was still mostly on nature writing. Once ASLE was established, in the early 1990s, and once *ISLE* appeared in 1993 and *The Ecocriticism Reader* in 1996, the situation rapidly changed. The kind of writing we associate with Thoreau and Annie Dillard was quickly complemented by critical writings that were more self-conscious about their own genres. In *The Ecocriticism Reader* itself, some of the most influential selections dealt with religion, with toxic consciousness in the American novel, with eco-feminism, Native Americans, technology, and psychology. Environmental justice, the "nature" of cities, environmental law, evolution, biology, and sociobiology produced further expansions. Although one hears occasional complaints from critics such as the above-quoted editor from Cambridge about the insufficient theorization of ecocriticism and its failure to nail itself down into a well-defined theory of critique, this weakness, if it is a weakness, has a compensatory strength. Ecocriticism covers a broad spectrum. It serves as an umbrella term for very diverse activities, and it is unwilling to be hijacked by a single, theoretical, fetishistic perspective.

My own view is that unlike literary Marxism, feminism, and queer theory, ecocriticism will be around for a long time, able to shape itself to unanticipated developments in the arts, society, and even international relations in an age of global warming. As Lawrence Buell put it in *The Future of Environmental Criticism,* "A telltale index is the growth within the last decade of the Association for the Study of Literature and Environment (ASLE) from a localized North American ferment into a thousand-member organization with chapters worldwide from the UK to Japan and Korea to Australia–New Zealand. The 'Who's listening?' question which nagged me when I first entered the arena of environmental criticism has given way to 'How can I keep pace with all this new work?'"[3]

The essays and articles that I produced after the initial eco-trio reveal some of the enlargements of ecocriticism I have pointed toward. They were not teased out by any theories but arose from an intuitive sense of the complexity of environmental matters in their social, philosophical, scientific, and political aspects. Like so much religious piety that is in fact a form of higher narcissism (e.g., the Big Bang took place 14 billion years ago in order to "save" my precious little "soul" today), "deep ecology" struck me as a painfully self-deceptive form of anthropocentrism. In the conflict between Dave Foreman and Murray Bookchin in chapter 6, my sympathies are clearly with Bookchin's attack on Foreman's over-the-top biocentrism. And Foreman himself began to moderate his positions as time went by, toning down the misanthropy. In dealing with Aldo Leopold (in chapter 7), I wanted both to represent him as an eco-saint while also pointing out his feet of clay: his biocentrism was not exactly what he thought it was, nor what many of his readers supposed. About Lawrence Buell's *Environmental Imagination* I had decidedly mixed feelings, as indeed Buell himself now does when he refers to it in *The Future of Environmental Criticism*. And in "The 'Environment' Is Us" (chapter 9) the quotes around "environment" are an early giveaway of my growing sense that there really is no environment, but that idea is not developed until much later (see chapter 17, "Ecocriticism's Big Bang").

In what is to date my favorite ecocritical insight, "Ecology and Ecstasy on Interstate 80" (chapter 10) turned out to be a contrapuntal weaving of environment, esthetics, and technology. Far from being a Luddite or indulging in Heideggerian moonshine, I acknowledge the ineluctable fact that even the most rarefied "spirituality" is beholden to technology and that almost everything human is enabled by what the Greeks called *technē*. Aldo Leopold is again a begetter in "Full Stomach Wilderness and the Suburban Esthetic" (chapter 11), where I use his full-stomach remark as an epigraph for another specimen of heterodoxy in which I acknowledge that even suburban expansion and destruction of the "wilderness" provide access to and appreciation of the very wilderness it partially destroys. And finally, for the present book at least, I try to unravel the ambiguities of J. M. Coetzee's ecological fictions.

Inevitably, however, ecology has to be seen as a component of Darwinian evolution and selection, the focus of part 2, and equally inevitable is the movement into consciousness studies in part 3. If there really is no environment, just morphing materiality, then natural selection goes a long way toward producing the sought-after Grand Theory, and consciousness, like trees and pollution, is just another one of its material products.

CHAPTER SIX

Ecology and Ideology

A MONG THE MANY VARIANT HISTORIES of Western culture that could be produced, a social-psychological account of the images and figures created by human beings to represent themselves and their desires in a flattering light would not be a waste of time. The self-aggrandizing reality behind the spiritual pretensions of "image of God," the normatives of "Reason" in the eighteenth century and "Nature" in the nineteenth, the Marxian figure of a "Realm of Freedom" (a secular heaven to be automatically arrived at once the proletarian revolution reached its successful conclusion)—these are just some of the most well-known tropes in mankind's ideological cornucopia.

Today, with so many ideologies in ruins, a new intellectual universe known as "ecology" has emerged from the residues of these outmoded systems of belief, sometimes revitalizing inspired insights, while at other times just recycling old and benighted mistakes. As ecology has moved with urgency into the higher consciousness of Western societies, a consciousness that expresses itself in such everyday forms as energy conservation, recycling, fear of chemical and nuclear disasters, and concern about deforestation, the ozone layer, and the greenhouse effect, a restoration of the age-old awareness of people's connectedness with the material that produced them (i.e., the earth) has begun to take place after a long moratorium during which the Industrial Revolution made mankind seem self-creating, autonomous, and omnipotent. This awareness has permeated not just the sciences—in the form of alternate-energy engineering, plant pathology, the chemistry of waste disposal and recycling, genetic studies of toxic mutations, environmental medicine, etc.—but the humanities as well. Less known to the educated general reader are the myriad ways in which ecology has filtered through philosophy, ethics, sociology, political science, psychology, history, economics, legal studies, religion, and, even more surprisingly, literature and literary criticism. A sense of this development can readily be obtained from a sampling of titles of specialist journals born in the past twenty years: *Environmental Ethics, Environmental History Review, Capitalism/Nature/Socialism, Earth First!*; these give only the merest hint of the intellectual ferment generated by ecological awareness since the end of World War II.

What is even less popularly perceived is the extent to which ecological activities are not simply free-floating and disconnected acts to "stop pollution" (let us say) but products of larger belief systems (both examined and unexamined) that aim to define the nature, value, and ends of life on this planet. These include such programs as "deep ecology," "eco-feminism," "social ecology," to cite only the most ambitious. In a word, ecology is also philosophy, politics, theology—or to use a term more congenial to today's intelligentsia, ideology.

At the high end of the ecological conversation, we are confronted with several antinomies that roughly define the specialized subsets of the discourse: deep vs. shallow ecology, biocentrism vs. anthropocentrism, reform vs. radical ecology; and there are myriad terms that shape and shade the major positions, such as conservationist, preservationist, liberal, leftist, bioregional, New Age, mainstream, and so forth. The temptation to place some of these terms in quotation marks is very great, since there is a tendentious, self-important, polemical, sometimes hoaxingly "metaphysical" quality about them. A lucid mini-survey of this social/philosophical ecology scene (and its bibliography is already daunting) is supplied by Steve Chase in a brief introduction to one of three remarkably interesting books that exemplify and participate in the heated debate going on within radical ecology: Dave Foreman's *Confessions of an Eco-Warrior*, Murray Bookchin's *Remaking Society*, and, with Chase's introduction, a "dialogue" between Foreman and Bookchin, *Defending the Earth*.[1]

But a few key terms need clarification. For example, although "reform ecology" sounds good, it is often used disparagingly to refer to what its enemies see as mere tinkering with the dominant capitalist, expansionist, human-oriented, wasteful Western mode of life that is destroying the planet. For some activists (like Dave Foreman), "reform" equals the "shallow" ecology of conventional left-wing politics. Organizations that many people would be inclined to admire, such as the Natural Resources Defense Council, Environmental Action, the League of Conservation Voters, and the Sierra Club, are dismissed by radicals as co-opted tools of industrialism and technology, contented to wear suits and ties and to imitate the legalistic maneuverings of their bureaucratic oppressors simply in order to change the route and speed—but not the overall direction—toward eco-hell. "Radical ecology" (or deep ecology), on the other hand, wants to abandon, in varying degrees, the basic institutions—economic, political, even spiritual—that characterize Western society. Reformists are seen by radicals as "anthropocentric" rather than "biocentric" or "ecocentric." They regard the planet (radicals claim) as made for the use and pleasure of human beings regardless of the cost to other forms of life and matter that occupy the earth. Conversely, radical biocentrists find "intrinsic value" in wilderness, grizzly bears, prairies, riverbeds, and even stones, though they are apt to waffle about rats and the AIDS virus.

The aboriginal antinomy of contemporary ecology, conservation/preservation, has been described by Roderick Nash in his classic work, *Wilderness and the American Mind,* as a schism that "ran between those who defined conservation as the wise use or planned development of resources and those who have been termed preservationists, with their rejection of utilitarianism and advocacy of nature unaltered by man. Juxtaposing the needs of civilization with the spiritual and esthetic value of wilderness, the conservation issue extended the old dialogue between pioneers [who cut down the forests as they destroyed the westward-moving frontier] and Romantics [who found God and beauty in nature unspoiled]."[2] "Wise use" and "conservation" are taken by biocentrists to mean a kind of benignly exploitative anthropocentrism, a planned destruction of "resources" for the benefit of Joe Sixpack hunters, motorized tourists, and middle-class human beings in general (with Theodore Roosevelt as progenitor). For biocentrists, "resources" is a dirty word, suggesting that the planet exists as a hoard of raw materials for the welfare of bourgeois mankind.

Viewed from a distance, the extremes of ecological Right and Left look something like—on the one hand—a patriarchal George Bush (père), willing to trash wetlands, air quality, wilderness, energy conservation, and even people (if they are poor, Iraqi, or Anita Hill) in the interests of ruthless capitalist expansion; and—on the other hand—animal-rights terrorists who bomb laboratories and emancipate research animals; or Earth First! ecoteurs, who spike old-growth trees (to make them dangerous to fell with chainsaws) and sabotage "development," in the interests of "intrinsic value." (The story is complex, but an excellent, if tendentious, history of these movements is given in Roderick Nash's more recent book, *The Rights of Nature*.) The ideological fringes become hard to sort out, however, for a legitimate question is why animals' rights and fetuses' rights fall at opposite ends of the political spectrum instead of in the very same position.

Dave Foreman and Murray Bookchin focus very bright light on one area of the *ecological* spectrum by means of a trio of vigorously programmatic books that register the quarrel between two "radical" environmentalists who inhabit different social, philosophical, and political worlds while intersecting uneasily at a number of key points. The jacket of *Confessions of an Eco-Warrior* pictures a rough and ready, burly and forty-fivish Dave Foreman in an Earth First! T-shirt. The impression conveyed is more accurate than not, for although Foreman is a reflective person who can proffer some reasonable ideas when his mood is right, his frequently antinomian thinking is driven by the powerful emotions of an activist bent on the preservation of wilderness and biological diversity. Because the book collects essays on ecology written between 1980 and 1991 (the exact provenance and alterations are left unclear), there is a certain amount of growth and development, as well as inconsistency, for which Foreman makes a non-apologetic apology.

The book opens in Foreman's most militant, absolutist, and gut-driven vein but increasingly softens as a result of criticism, FBI persecution, re-reflection, socialization, and disillusionment, so that by the time we reach the end, we are not quite sure who this "Dave Foreman" person really is. Is he a grizzly bear with a teddy bear's heart, or a teddy bear with a grizzly bear's heart? It's anybody's guess, because one of Foreman's most striking characteristics is his unremittingly revisionary narration of his own personal history in ecology, which he rewrites like crazy as each new event or philosophic encounter shakes up his universe yet again. (Confronted in debate with Murray Bookchin in *Defending the Earth*, his rewriting machine seems on the verge of breaking down altogether from dangerously high rpms.)

Foreman worked for the Wilderness Society from 1973 to 1980 but was gradually disaffected by its capitulations to the growing mainstream American environmental movement. Tired of compromises with federal bureaucracy, in 1980 he became one of the founders of Earth First!, a more or less "apolitical" action group that employed direct intervention (like throwing themselves in front of bulldozers) to stop environmentally destructive development, deforestation, and dam building as well as road construction in wilderness areas. In the early chapters of his book, Foreman is at his most antinomianly biocentric, offering us New Age style inspiration as counterpoise to the entire post-Enlightenment Western humanist tradition, which he claims to detest (for its reliance on instrumental—i.e., exploitative—reason). This takes a lot of chutzpah and would seem doomed to failure, even if it *can* claim Aldo Leopold's "land ethic" as its father (an ethic which, to quote Leopold, "changes the role of *Homo sapiens* from conqueror of the land community to plain member and citizen of it"). Or as Foreman himself puts it:

> Human beings are merely one of the millions of species that have been shaped by the process of evolution for three and a half billion years. According to this view, all living beings have the same right to be here . . . A Grizzly Bear snuffling along Pelican Creek in Yellowstone National Park with her two cubs has just as much right to life as any human has, and is far more important ecologically. All things have intrinsic value, inherent worth. Their value is not determined by what they will ring up on the cash register of the gross national product, or by whether or not they are *good*. They are good because they exist.
>
> Even more important than the individual wild creature is the wild community—the wilderness, the stream of life unimpeded by human manipulation. (3–4)

Borrowing the technique of ecological sabotage called "monkeywrenching" from Edward Abbey's novel *The Monkeywrench Gang,* Earth First! has achieved a mixed record. Not so much by saving ecosystems directly as by obtaining media attention that ultimately influenced pro-environmental legislative and judicial decisions, Earth First! made the mainstream environmental organizations (such as the Sierra Club and Natural Resources Defense Council) look more temperate and reasonable so that federal and state governments began to pay them more heed. But on the negative side, Earth First! could seem a motley band of ragtag, prankish Yippies, or misanthropic, destructive, deranged true believers. Foreman spends much of his book outlining the group's principles and accomplishments and, it must be admitted, his revisionary account makes them appear, on balance, a positive force. "We do not engage in radical action because we are primarily motivated by opposition to authority, because we are antinomians, but because we are *for* something—the beauty, wisdom, and abundance of this living planet" (214). But as in most of Foreman's historicizings, it is hard to tell how much is rewriting to suit the temper of the times and how much is accurate reportage.

Biocentrism, whether that of Foreman or anyone else, is a position fraught with self-contradictions and absurdities, since all life feeds on other life, and no life *except the human* has any regard for the preservation of anything but itself. Biocentrists eat, drive cars, use paper, step on ants, kill bodily invaders with antibiotics, and all the rest of it; so their position, at best, becomes a question of degree that only leads to trivial side issues, such as whose self-indulgences are the least harmful or the most virtuous—and volumes could be written on the meaning of "harmful" alone. (The only complete biocentrist would be a dead one, returning his elements to the earth for recycling.)

The relentless need for greater and greater economic output that characterizes capitalist-industrialist Western society, a need that is destroying the planet and subverting the precise mix of elements required for life to continue, is not circumvented or even seriously addressed by extolling, as Foreman does, the lives of hunter-gatherers as "healthier, happier, and more secure than our lives today as peasants, industrial workers, or business executives" (28), or by insisting, as many ecologists do, that the early European settlers of America supplanted a native lifestyle of golden-age qualities with a vicious and destructive one. That American Indians thanked the deer and fish they killed for dinner may have been courteous (though dead is dead, whether thanked or unthanked), but it has little bearing upon any substantive biocentric virtue that would rescue us from our own high-tech depredations (and if Native Americans had been allowed to multiply instead of being murdered, no amount of courtesy would have solved their own depletion of natural resources as they grew into a population as large as India's or China's or our own). While Foreman's desire for a drastically decreased population makes good sense, welcoming the AIDS virus or starvation in Ethiopia (remarks that Foreman claims have been taken out of context in order to bad-mouth him, but of which he is not entirely innocent), or referring to humanity as "the human-pox," only contribute to one's sense that underlying the self-satisfied virtue of biocentrism is a misanthropic elitism that regards "nature" as an esthetic object that ought to be preserved for the pleasure of finer sensibilities.

Like Aldo Leopold himself, whose highly esthetic appreciation of flora and fauna coexisted with a love of hunting and disdain for the rabble, Foreman admits without shame to being a hunter who prefers the taste of elk, contrasting himself with the mass-mind "slob" hunter-murderer equipped with vulgar high-tech road vehicles, who shoots anything in sight, or "the thrill-seeking dirt bikers who terrify wildlife and scar delicate watersheds with mindless play" (124), or the suburban family that wears fur coats or thrives on antiseptically packaged, factory-produced supermarket meat that has already been killed for their convenience. Yet dead would still be dead, no matter how exquisite the tastebuds (or refined the sensibility) of the hunter-killer who paradoxically insists that all life has an equal right to exist.

Foreman's aversion to the politics of both the Left and the Right is implicated in his belief that economic and social problems are less important than ecological ones and that today's bureaucratic environmentalists "must be guided by the vision of Muir, [Rosalie] Edge, [Robert] Marshall, and Leopold—not by that of the Harvard Business School" (212). In light of the evils perpetuated by capitalist expansion, this may sound plausible, but it is a little like saying that the United States should be run according to insights derived from Wordsworth's *The Prelude* rather than by the U.S. Constitution. It is the rule of law rather than personal inspiration that has protected us so far from the horrors of Stalinism and Hitlerism. Without it, we would be totally, rather than partially, at the mercy of the Bush-Specter-Simpson-Hatch *kick their ass, gun 'em down* mindset.

What Foreman doesn't intuit is that ecology, even his own purist version, is one of the "human sciences," *bestowing* value on "nature" as all consciousness bestows the value it pretends to find already given in its own local culture. The "intrinsic worth" that biocentrists connect with animals, plants, and minerals is projected by the desiring human psyche in the same way that "the will of God" is projected by human vanity upon a silent universe that never says anything, let alone anything deemed valuable by all human cultures. The "biocentric" notion of "intrinsic worth" is even more narcissistically "anthropocentric" than ordinary self-interest because it hopes to achieve its ends by denying that oneself is the puppeteer-ventriloquist behind the world one perceives as valuable. But when Foreman is driven to ask whether the real world is in some way connected with our heads, his reply is "No! The real world is out there—independent, autonomous, sovereign, not ruled by human awareness. The real Grizzly is not in our heads, she is in the Big Outside—rooting, snuffling, roaming, living, *perceiving on her own*" (51). For him, that a certain person named Dave Foreman happens to have a fix on grizzlies (as others of us do on artworks, cars, or kinky sex) has nothing to do with human subjectivity and its attribution of value. Foreman talks about natural beauty as though it were a self-presenting absolute, rather than a post-Enlightenment invention that happened to require subduing the wilderness and becoming bourgeois to appreciate. Indeed, Aldo Leopold himself remarked, "These wild things, I admit, had little human value until mechanization assured us of a good breakfast, and until science disclosed the drama of where they come from and how they live," and Roderick Nash refers to nature appreciation as "a full stomach phenomenon." There are still some cultures, as I keep reading, that do not even understand the conception of "wilderness"—which we now highly value as it disappears—because they live so completely *within* a wilderness that they have no non-wilderness with which to contrast it. "Wilderness" is *our* creation, not God's.

As his book proceeds apace, a more measured and public language begins to mute (but never eliminate) the solipsism of Foreman's biocentrism. The most rewarding chapters involve extremely knowledgeable accounts of the concrete facts behind the disappearance of America's forests and wild animals. As a hands-on environmentalist, Foreman has experienced for himself just about every habitat and ecosystem in the United States and is unrivalled in his authoritativeness about their nature and decline. His recitation of the facts is more alarming than any of the millennial outcries of his earlier pages. Even his radical view that a great deal of public land already invaded by roads and "improvements" should be returned to wilderness has a lot going for it. (The Sunday *New York Times* of November 3, 1991, devoted almost two pages to the mismanagements of the U.S. Forest Service and the depredations of clear-cutting old growth forests.)

By the last quarter of the book, in an act of supreme rewriting, Foreman expresses the awareness that even intensely held beliefs like his own are myths: "My mythology and that of my associates is Deep Ecology, or biocentrism. But no matter how valid it is, how deep it is, we must constantly acknowledge that it is still an abstraction. It is a good, workable basis by which to operate. But it is not infallible scripture" (175). Amen.

Murray Bookchin, as passionate in his way as Foreman, comes from an altogether different tradition. If Foreman's sensibility is saturated with the values of the Southwest (he was born in Albuquerque), Bookchin is a New York intellectual in the down-to-earth yet utopian tradition of Paul Goodman. With a prolific output of books, he has been a ceaseless social activist (for more than fifty-five years, he reports)—latterly in Vermont with the Vermont Greens and the Institute for Social Ecology, which he cofounded in 1974, not to mention his professorial identity at Ramapo College in New Jersey.[3] Given his extensive output and a magnum opus, *The Ecology of Freedom,* that is a dense and taxing work to get through (though Bookchin always writes well), he has made good on the suggestions of his friends that he produce a concise and accessible distillation of a lifetime's thinking. *Remaking Society: Pathways to a Green Future* is Bookchin at his best.

Bookchin's trademark term is "social ecology," with its distinctive doctrines, anarchy and anti-hierarchalism. In their working out, these ideas reveal that there are *two* Murray Bookchins: the brilliant, mordant, racy, realistic social philosopher and the pie-in-the-sky utopian idealist. The utopist focuses on the past and the future, the realist on the present. This past is a mythic preliterate golden age (based on very questionable anthropological assumptions derived from fairly flimsy archaeological evidence) in which a nonhierarchical and non-dominating family of hunter-gatherers lives harmoniously, unmenaced by male supremacy and patriarchy, each sex having its roles and chores but with no sense of superior, inferior, or dominant-servile. Land is shared and "self" unknown. "It is hard for the modern mind to appreciate that precapitalist societies identified social excellence with cooperation rather than competition; disaccumulation rather than accumulation; public service rather than private interest; the giving of gifts rather than the sale of commodities; and care and mutual aid rather than profit and rivalry" (46–47).

Bookchin choreographs the falling off of this primeval egalitarian society through stages involving the gradual acquisition of power, first by elders, then by males, warriors, small communities, city-states (e.g., the Athenian *polis*) with direct representation (but slavery), and finally nation-states with indirect (or no) representation, dominated by capitalist or communist power structures. This increase of hierarchy, domination, and exploitation of each other is part and parcel of our domination of nature; in other words, it is hierarchy not anthropocentrism that underwrites today's ecological catastrophe. But Bookchin realizes, "There is no way that we can return to the naive egalitarianism of the preliterate world or to the democratic *polis* of classical antiquity. Nor should we want to do so. Atavism, primitivism, and attempts to recapture a distant world with drums, rattles, contrived rituals, and chants whose repetition and fantasies bring a supernatural presence into our midst" (71)—that is, some of today's New Age earth-mother and nature-worshiping ecological regressions— won't cure our ills but merely add to them through devaluation of the human and destruction of hard-won social institutions.

What Bookchin means by "social ecology" is that ecological problems are only one aspect of a system of social domination extending from the human being/nature relationship to the person/person relationship in which both people and nature are degraded into the role of "resources" for exploitation by those on top. Ecology thus becomes a *social* science that finds hierarchical power relations at the heart of all forms of exploitation and oppression.

In the course of tracing the movement from an imagined ideal past to an imagined ideal future, Bookchin exhibits his skill as one of the most intellectually powerful critics of contemporary society, one who sees through the cant, self-deception, and destruction involved in capitalism, Marxism, and (lamentably) certain varieties of ecology. The ruthless expansionist premise of capitalism requires that more and more forests, minerals, animals, *and people* be used to produce more and more products, many of them unnecessary, in order to keep the economy in constant growth. As a result, people are forced by exploitative and mendacious corporations—whose operating principle is "grow or die"—into humanly and ecologically destructive roles as workers and consumers. Like Foreman, Bookchin believes that any attempts to "green" capitalism are doomed, but whereas Foreman somehow advocates a retreat to wilderness, without political action—he recently left Earth First! because he saw it being taken over by leftist ideologues—Bookchin is an old-time leftist-activist (who detests Marxism for its reduction of humanity to mindless, class-defined automata whose essence is to labor).

The big clash between Bookchin and Foreman that ended up as the public debate recorded in *Defending the Earth* came about in the late eighties as a result of Bookchin's vehement denunciation of biocentrism. Whereas Foreman attacks humanism and the Enlightenment for the anthropocentric priority they give to "reason" and human interests (as opposed to the rest of creation), Bookchin defends Enlightenment reason (from the onslaughts of eco-feminists and wilderness freaks) as having "brought the human mind from heaven down to earth . . . It fostered a clear-eyed secular view toward the dark mythic world that festered in feudalism, religion, and royal despotism" (166). Furthermore, he adds, "the abuse of these ideals by industrial capitalism through the commodification and mechanization of the world does not invalidate these ideals by one whit" (166). Since the complete abandonment of Enlightenment ideals (advocated by various extremist crazies) would entail a return to antinomian savagery (and Foreman himself, as well as some eco-feminist, animal rights, and pro-life crusaders, at times displays this primitivist, misanthropic strain), Bookchin believes that "only a dialectic that combines searching critique with social creativity can disassemble the best materials from our shattered world and bring them to the service of remaking a new one" (170–71).

Seeing deep ecology as "spawned among well-to-do people who have been raised on a spiritual diet of Eastern cults mixed with Hollywood and Disneyland fantasies" (11), Bookchin is scandalized by their equation of human beings with trees and stones and their reversion to primitive nature-oriented religions. He continually emphasizes that evolution has resulted in the greater and greater ability of species to direct their own subsequent development, "nor will ecological creativity be served by dropping on all fours and baying at the moon like coyotes or wolves. Human beings, no less a product of natural evolution than other mammals, have definitively entered the social world. By their very own biologically rooted mental power, they are literally *constituted* by evolution to intervene into the biosphere" (71–72).

Bookchin's "ecological society" of the future would replace capitalism with a nonhierarchical "libertarian municipalism," decentralized in small towns and cities through a confederation of local governments, with direct, face-to-face representation of the population. These communities would be tailored to the local ecosystems, acquiring an energy and commodity self-sufficiency that would eventually eliminate the use of non-renewable fossil fuels, the extreme waste involved in shipping food and other products over long distances, and the domination over people and nature by the top levels of a power structure caught in a "grow or die" mentality. "Either we will turn to seemingly 'utopian' solutions based on decentralization, a new equilibrium with nature, and the harmonization of social relations, or we face the very real subversion of the material and natural basis for human life on the planet" (185).

Although all three of these books are the distinctive creations of powerful personalities, Bookchin's *Remaking Society* has the aura of a classic that deserves to be treated as an essential text of social philosophy, while *Defending the Earth* provides general readers with a compressed statement, in their own voices, of *both* Foreman's and Bookchin's ideologies. Although the two debaters exhibited tact and conciliation in their public interchange (with a certain amount of Alphonsing and Gastoning), they adhered nonetheless to their wonted doctrines, creating the taut atmosphere of two lions trying hard not to bite. But in a pair of appended "Closing Essays" written a year after the public event, both spokesmen return to more uncompromising restatements of their views.

The final question, of course, is how seriously one can take Bookchin's utopian recommendations for bringing about an "ecological society." While Foreman's wilderness doctrines seem to be floating in a void of undefined social and political contexts, so that one does not understand how they mesh with any existing or projected world, Bookchin has more concretely spelled out a context for the implementation of social ecology. Both Foreman and Bookchin are certainly right about capitalism's failure or inability to deal with ecological problems until they become dire. The GNP must be expanded until the *Pequod* goes down into the briny deep. New attempts to drill oil in Alaska or to redefine wetlands, the latest plans for reviving nuclear power, the conflict between loggers and the spotted owl—will these assaults upon our own particular Moby Dick continue on to the bitter end, with Chernobyl heaped upon Chernobyl, war in the Gulf piled upon war in Central America? Must serious ecological reform be resisted as intractably as a market economy in Russia, because avoiding short-term pain takes precedence over the likelihood of more distant apocalypse?

Though Bookchin's far-fetched ecotopia will never voluntarily come to pass, one is gripped by the underlying truth of its scenario and a fear that everything will suddenly come crashing down all about us (or our children), necessitating reconstructions even more drastic than he now envisions. Today it's only a sudden ozone hole over North America, unanticipated by prescient experts who saw the immediate problem as principally polar. Or it's a few thousand drownings in the Philippines as a result of rapacious and illegal clear-cutting of mountainside forests that formerly absorbed floods. Will tomorrow's catastrophic surprises make today's cry for wilderness seem like a trifling, decadent luxury?

CHAPTER SEVEN

Aldo Leopold

Esthetic "Anthropocentrist"

ALL LIFE CONTINUES IN EXISTENCE by feeding on other life, favoring itself at the expense of everything else. Though crude, depressing, insane, no way has yet been found to circumvent this enabling murderousness—except by means of upbeat redescriptions, like "image of God," "realm of freedom," "new world order." Thus, mice, rats, cockroaches, and the AIDS virus look to their own survival at all costs, and people are necessarily anthropocentric. Biocentrism, a recent invention that one might call "cosmic pro-lifeism," entails the redescribed alter egos of certain types of well-fed, bourgeois anthropocentrists, more or less freed from the struggle for survival, and now with time on their hands for romancing the wild from which they have been emancipated by the technology that keeps them alive with little effort, but which they frequently profess to hate. Indeed, Aldo Leopold, a pretty straight talker, in his introduction to *A Sand County Almanac* says of "wild things": "These wild things, I admit, had little human value until mechanization assured us of a good breakfast."[1]

But do real biocentrists eat breakfast, or anything at all? For the authentic inaugurating act of a would-be biocentrist should properly consist of suicide, since by staying alive he uses up another creature's resources— even its very life. To be alive, it would seem, is to be against life, or at least everyone else's life except one's own. But nobody appears to be doing themselves in out of biocentric remorse. On the contrary, "biocentrists" consume paper, electricity, computer products, as well as food (and jet fuel to attend conferences) just like ordinary people (and jet fuel could be said to have been made available not just by the ancient deaths of fossils but by the recent deaths of hundreds of thousands of Iraqis). Biocentrism begins to look a lot like one more redescription of the anthropocentric will to power, with an agenda whose worldly underpinnings are conveniently muffled by transcendental neologisms. In a word, a lot like conventional religiosity. Instead of the "will of God," one invokes the "will of the biota."

Still, even Roderick Nash in *The Rights of Nature* is ready to concede that "no environmentalist seeks literal equality for the subjects of his or her concern,"[2] but such a concession (that one may perhaps be a little bit anthropocentric) is tantamount to admitting that one may be only a little bit pregnant. Because once a "biocentrist" is free to pick and choose, there is little to distinguish him from vulgar anthropocentrists, who also pick and choose, since very few people are total monsters of depravity. Though I myself sometimes step on ants and take antibiotics to murder bodily invaders, I certainly don't ever step on dogs and cats. What must I *really* not step on to qualify as a bona fide biocentrist? The toes of other biocentrists? But I shouldn't say "other biocentrists," because like Aldo Leopold, I'm just an anthropocentrist.

Why then did I put ironic quotation marks around "anthropocentrist" in referring to Leopold in the title of this paper? Not because Leopold is not really an anthropocentrist. But because everybody is an anthropocentrist, except corpses pushing up daisies: *they* are the real biocentrists, giving their all so others can live. When John Muir talks about "thinking like a glacier," or when Leopold talks about "thinking like a mountain," they are engaging in quintessentially anthropocentric appropriations of reality, for to think like glaciers or mountains is already to have nothing to do with those things and everything to do with people. Only a person can think like a mountain, and that thinker is inevitably someone whose genetic inheritance is to think like an anthropos, never more thoroughly than when he is "thinking like a mountain." Biocentric terms like "ecological egalitarianism," "inherent value," "a sense of place," "bioregionalism," "ecosystem," "sacred space," "aesthetic experience of the wilderness," "caring about nature" (all of which I've taken from Devall and Sessions' *Deep Ecology*)[3] are saturated through and through with the anthropocentrism of creatures constructed like us. To attempt to think "biocentrically" is to try to sneak a look through the back door of the universe so quickly that one's observations would escape the indeterminacy principle and one would see things as they really are in their unseen selves. But things as they really are in their unseen selves are presumably not perceptions or thoughts. No matter how empathetically we try to apprehend noumena on the sly, the act of knowing in itself transforms them into phenomena, that is, into humanized interests. If it is therefore impossible for human beings to know the intrinsic interests of animals and trees (because knowing is the quintessential anthropocentric act of appropriation), perhaps when we talk about the interests of trees we are really talking about our own interests, as when we used to talk about the will of God.

This is not to say that there is no difference between selfishness and unselfishness, between inhumaneness and humaneness. Rather, even unselfishness (call it "biocentrism" if you wish) derives its force from a context of human interests in which neither trees nor animals participate. Yet despite this interestedness, only human beings have displayed the faculty of empathy with the rest of creation, an empathy entertained by no other species, however much it is a projection of human pathos upon unknowable "Others."

The paternity behind much of today's rights-based and deep-ecological ethics is Aldo Leopold's pioneering work, *A Sand County Almanac,* written over the course of many years before being published posthumously in 1949. Since this book has now achieved almost scriptural status, a brief but revisionary glance at its purported biocentrism is needed in order to correct what has latterly become an out-of-context misappropriation of a few germinal sentences from the section called "The Land Ethic."

In this by-now excessively quoted chapter, Leopold introduces (or reintroduces) for his contemporaries the idea that the use of the earth solely as an economic resource will eventually destroy both it and us. Ethics, therefore, must be extended to include "soils, waters, plants, and animals," and humans must change their role from "conqueror of the land-community to plain member and citizen of it" (204). His most cited statement is, "A thing is right when it tends to preserve the integrity, stability, and beauty of the biotic community. It is wrong when it tends otherwise" (224–25). These remarks, which are made in the course of a rich and well-considered account of contemporary ecological deterioration (and things were much less dire when Leopold wrote than they are now) have been taken up as part of a new set of doctrinal imperatives by a number of recent biocentric ecologists. Leopold's aim, however, was to show the extent to which society's response to nature had been determined almost exclusively by economic considerations throughout the colonial and postcolonial periods of United States history. To redress this imbalance, therefore, he warns us: "(1) That land is not merely soil. (2) That the native plants and animals kept the energy circuit open; others may or may not. (3) That man-made changes are of a different order than evolutionary changes, and have effects more comprehensive than is intended or foreseen" (218).

After outlining the character of "the land pyramid" and the operation of its food chain in order to suggest this comprehensiveness, he concludes with a summary: "A land ethic, then, reflects the existence of an ecological conscience, and this in turn reflects a conviction of individual [as opposed to purely governmental] responsibility for the health of the land. Health is the capacity of the land for self-renewal. Conservation is our effort to understand and preserve this capacity" (221). At the time he makes his famous remark about "integrity, stability, and beauty," he has been urging his readers to take into account not just economics (which he concedes will always be foremost) but "what is ethically and esthetically right" as well. In other words, trying to counterbalance the overwhelming force of almost universal ecological short-sightedness in the 1930s and '40s, he allows himself a moment of dogmatic insistence on the longer perspective.

But Leopold's now almost Mosaic criteria, far from being inscribed on sacred tablets derived from the biota itself, are rooted in ultimately anthropocentric concepts that have been newly refurbished by environmental proselytes to serve as "revelatory" foundations for a contemporary eco-theology. Taken as absolutes lifted from the needs of Leopold's rhetorical context, however, these criteria pose serious problems. The notion of "wholeness" or "integrity," for example, has come in for a good deal of post-structuralist criticism, particularly in connection with the old "New Criticism's" touchstone of "organic unity," but it is also generally dismissed in other fields besides the literary. Understood to exist in the mind of the beholder, who selects a number of qualities and data to stand for the whole while ignoring everything else, integrity or wholeness are nowadays seen as purely conventional moments of understanding, not aspects of "reality." Could anyone ever expect to enumerate all the possible qualities and data that might be said to inhere in any given entity or system? Indeed, to name them is in large measure to create them, since colors, textures, relationships, etc., are mind-dependent. And if an entity's essential characteristics cannot be finitely identified, how can anything be proclaimed to be a system or whole? ("O chestnut tree, great rooted blossomer," asked Yeats, "Are you the leaf, the blossom or the bole?" And what about the chemical transactions of symbiotic microorganisms?)

Thus, the qualities and data involved in describing a system would appear to have little to do with "nature" and a lot to do with the cultural history and teleological interests of the describer. As for "stability," the belief that ecosystems are stable is no longer generally supportable. Daniel Botkin, having devoted an entire book to demonstrating the falsity of this idea, explains how stability, in a case like that of ecosystems, is attuned to human perceptions of what is relatively enduring (from a short-term perspective) in a constantly changing material universe. "Wherever we seek to find constancy we discover change,"[4] a phenomenon that Botkin illustrates over and over again in his discussions of forests, predators and prey, winds, fire, elephant preserves, and birds. By the end of his book, the idea of "nature undisturbed" seems like an incomprehensible contradiction in terms. As for "beauty," it is too obviously culturally determined and consciousness-generated to require comment. (John Passmore points out that wild alpine landscapes were regarded as junk vegetation before the late eighteenth century, and owe their esthetic appeal to the cultural program of the Romantics.) Perhaps speaking in a figural way one could say Leopold's outlook is "biocentric" as compared with the traditional attitudes that he criticizes, but when his entire book is taken into account, Leopold's preoccupations look simply like another set of *human* interests, different from those of General Motors and Exxon, and almost certainly better *for the world of human beings* in the long run, but anthropocentric nonetheless.

Benign as *A Sand County Almanac* may be overall in its aim of preserving a usable and beautiful world, it has a regressive side as well. Leopold can at times appear to extol the preservation of "systems" at the cost of the individual members, with all of the transcendental religious implications that are present in such viewpoints. Although at first glance such positions may seem "biocentric" and "disinterested" in their apparent put-down of people, they can also be seen as a form of elitist, gnostic transcendentalism, related to Leopold's powerful response to natural phenomena, which he wants to preserve for esthetic contemplation and defend from the invading democratized rabble, with their motorcars, high-tech sports equipment, and sports columnists who tell them where the fish are biting. (It's not for nothing that the essays collected as *The River of the Mother of God* keep referring to "Mr. Babbitt," Sinclair Lewis's arch philistine, or to the mass mind, or to Ortega y Gasset's *The Revolt of the Masses*. Leopold, alive today, would never pass muster as a spokesman for political correctness.)

Indeed, despite his precursorship of today's "biocentrism," with its pretensions to cosmic egalitarianism, Leopold has no objection to killing for sport and can talk, just like you or me, about "worthless" grasses and vegetation. His "thinking like a mountain" is actually an expression of concern for the destruction of mountain vegetation by a deer population allowed to grow because human beings have killed their natural predators, the wolves. Because this denuding of mountains is in the long run harmful to various *human* interests, the esthetic as well as the ecological, he believes that predators must be allowed to flourish. Although most of Leopold's strictures would in fact benefit the human race at large, his own interest in them often betrays the concerns of an elite, high-toned sportsman with exquisite esthetic tastes verging on mysticism, though a mysticism sorely compromised by a powerfully atavistic (to use his own word) attachment to hunting that begins to trouble him only late in his life. (In his essay, "Goose Music," from the collection *Round River,* he rhapsodizes over the flights and sounds of geese while simultaneously extolling the pleasures of shooting them, a paradox he doesn't attempt to iron out.)

In sum, there is more ideological complexity and affective strife in Leopold's many-faceted book than is suggested by the handful of ecological imperatives that have been abstracted from it in the interests of postmodern biocentric politics. Indeed, the need to appropriate Leopold for what they prefer to call "noninstrumental" values drives even such philosophers as J. B. Callicott and Eugene Hargrove to criticize some readers of "The Land Ethic" for describing Leopold as anthropocentric, even though they concede such a reading is easily possible. They have jointly remarked that Leopold's program there and elsewhere is "primarily motivated by esthetic concerns, rather than concerns about human welfare. Thus [reading these writings] as grounded in instrumental rather than the intrinsic value of wild nature does not correctly represent Leopold's views as they historically developed."[5] But esthetic response is the most powerfully anthropocentric interest of all, produced as it is by the very nature and operation of our bodies and psyches: our metabolism, sense mechanisms, heart rate, sexuality, brain cells, and enculturation in temporal human societies. The "beauty of nature," strictly a "human interest" (however *indirectly* instrumental), leaves geese and bats quite cold.

Drawn to evolutionary biology to satisfy his frustrated religious longings, viewing the universe through esthetic glasses and thus disdainful of capitalism's cash nexus while at the same time acknowledging its inevitability (and its attractive side as well), caught somewhere between a down-home concern for the future of human life and a type of intellectualist snobbery, Leopold would very likely have a few wry words for today's Luddite, misanthropic biocentrists. Call him what you will, it is Leopold's highbrow anthropocentrism, with all its unresolved contradictions, that finally makes him so ambiguously attractive.

CHAPTER EIGHT

Postmodern Ecologizing

Circumference without a Center

EVERY NOW AND THEN a reviewer finds he is sorry to have undertaken to review a book that initially looked promising, because he knows that the outcome will probably make everyone unhappy. The reviewer will be unhappy because no matter what position he takes, he will be dissatisfied with the consequences. The author of the book will be unhappy because he is almost certain to feel he has been treated unfairly. And the readers of the review may very well be confounded by an emphasis on the book as a constructed artifact rather than as a vehicle for "contents."

I fear I am about to embark on just such a no-win exercise in trying to provide a fitting account of Lawrence Buell's *The Ecological Imagination: Thoreau, Nature Writing, and the Formation of American Culture,*[1] a book with a title more global and ambitious than its contents warrant. It has been given an enthusiastic launching by Harvard University Press and has been much bruited in ecocritical circles. At the very moment I had finished my reading and was despairing about how to handle it, I received in the mail from Harvard University Press a publicity sheet filled with puffs designed to make me feel even more rotten. My minimal consolation is that, apart from my own quirky response, I am sure the book will be well received, highly praised, and provide a generation of graduate students in American Studies (though generations are very brief these days) with plenty of material for dissertations, scholarly articles, and sessions for MLA conventions.

A hint of the problems to be faced occurs in the first paragraph of Buell's introduction: "This book has refused to remain the modest undertaking I intended it to be. Planned as a history of Thoreauvian writing about the American natural environment, it has led me into a broad study of environmental perception, the place of nature in the history of western thought, and the consequences for literary scholarship and indeed for humanistic thought in general of attempting to imagine a more 'ecocentric' way of being. I found that I could not discuss green writing without relating it to green thinking and green reading" (1). Having just read the book, I am forced to admit that I don't recall most of these immense aims being realized in any substantial way, though the subjects are more or less taken up at one point or another. The result, says Buell, "is an exploratory work with several foci rather than one," and that may be a large part of the problem. Buell attempts to rationalize this by saying, "The combination of broad sweep and cranky hyperfocus [on Thoreau] of which I have forewarned is, I think, in keeping with the nature of environmental representation, which is at least faintly present in most texts but salient in few." Whether "environmental representation" has a "nature" and whether this is it are open questions, but the book still seems unfocused, and even its key terms, like "environment" and "ecological," remain somewhat cloudy (quite apart from the fuzzy syntax of the sentence itself). When I turned the final page, I was left with the sense of a very learned ramble rather than "a broad study." Had the book's title and contents adhered to Buell's original plan, something like "Thoreau and Nature Writing in Nineteenth-Century America," the reader's expectations would likely have been more helpfully confined because that is where the major emphasis appears to be.

Before I report on a few of the book's strengths, I would prefer to address other matters first. Buell, a professor of English at Harvard, is probably one of the most learned of the Americanists now dominating the academic scene. His reading and interests, well represented by 137 pages of densely bibliographical notes (roughly one-quarter of the book), are not merely impressive, they are stunning. His acquaintance with world literature, the other arts, philosophy, history, and criticism is matched by the remarkable depth of his knowledge of American literary (and nonliterary) culture. Not only has he read the obscurest of American texts, "ancient" and modern, but his retention of their minute particularity is daunting. His prose is unusually allusive, echoing other writers on every page, and his acquaintance with contemporary culture—from high to pop—is up to the minute. He is so on top of just about everything that, perverse as it may seem, his mastery of the current moment of the flux of Western culture served for me as a *memento mori,* a precarious virtuoso athletic feat that called to mind the king of the woods—doomed to be displaced by a younger and stronger hero—that Frazer had so much to say about in *The Golden Bough,* a myth of transience, mortality, and eternal recurrence. The very brilliance of Buell's juggling act, his lifelong workaholism combined with a supersubtle sensibility and powerful intelligence, left me not only awed but somewhat depressed insofar as they bespoke both the desiderata and the futility of what nowadays constitutes an ideal academic career. In the back of my head I kept hearing Yeats's lines, "Everything that man esteems / Endures a moment or a day." To this, of course, one might reply, "So what?" Should the temporariness and impermanence of things preclude superhuman effort? Is the fact of mortality an argument for mediocrity? But things are more complicated than that.

Not only is Buell gifted with a magisterial intellect—calling to mind such polymaths as Harold Bloom and George Steiner—but he has an excellence of judgment and a scrupulous, rare sort of sanity and self-awareness far beyond the level of today's bemused race of frequently screwball academics. I don't recall him ever being taken in by anything nonsensical or merely trendy, nor is he conned by phony pieties, even when they are ecological. When he *is* sympathetic to politically correct shibboleths (and he generally is), he is still likely to express some reservation, some sense of balance, some residue of doubt. He is an intellectual, then, for whom I can personally acknowledge a great deal of respect.

If that's the case, then why all this twitching and dancing? Just what is my problem with *The Environmental Imagination*? I think my allusion to Harold Bloom and George Steiner could offer some assistance here (and I might as well throw in Camille Paglia [see chapter 14] for good measure). What differentiates these three polymath thinkers so sharply from Buell is that to one degree or another they're all a bit crazy. Their emotional, powerful, intemperate, intrepid, sometimes screws-loose writings are the products of overwhelming visions, *idées fixes,* strong moral convictions, sweeping prose styles, and imperial selves that make them appear to have some inner, profound, subterraneous connection with the universe that the rest of us wimpy mortals lack, however wrongheaded and ridiculous they all at times can be. The things they have to say alter our consciousness willy-nilly, even if we disagree with them and feel they've gone off the deep end. Yet at the same time, while I would be very amenable to following Buell's lead on most of the subjects he addresses, I would be quite wary of much that Bloom, Steiner, or Paglia had to say in the context of the everyday world. In sum, what I appear to be saying is that a certain madness (or "inspiration," if you prefer) is required to write a great book and that Buell is a virtuoso of sanity. Bloom, Steiner, and Paglia recognize that you *can't* make your sun stand still, so you've *got* to make him run. Buell is still hopeful that mortality can be outwitted by sheer effort, by piling Pelion upon Ossa, by carrying yet another coal to Newcastle.

Buell has one insight after another into the texts and writers he discusses, but he is like a metaphysician peeling off layer after layer of reality, ignoring the intimation that there are infinite layers for the peeling. (He is postmodern through and through. The religious call it "living without God," and for them it's bad. The Derrideans call it "living without a Center," and for them it's the only thing available.) If there *are* infinite layers, however, why take any layer in particular too seriously? A miss in this discourse is as good as a mile, if myriad layers still remain. It's the old story of the spider vs. the bee: the pragmatic bee sucks nectar from large numbers of flowers—and if he sucked nectar from just one more flower, the final honey would have a slightly different—and maybe better—taste. The spider, on the other hand, spins from her own substance (so to speak, since the substance ultimately derives from the earth). *Her* effect is a sort of inspirational *ipse dixit*. The irresolvable dilemma, then, is that clear-sightedness can always in theory see more (thereby weakening what it *does* see), whereas passionate conviction and vision are always at risk of being merely demented (see Bloom, Steiner, Paglia). Buell strikes me as obsessed with the notion of being unassailable: he reviews every possibility that occurs to him, avoids rashness, balances one extreme against the other. He knows that the academy is a snakepit and he has developed eyes all over his body to avoid sneak attacks that might accuse him of anything less than omniscience and "correctness." But as an author Buell is too self-aware for his own good. He's a victim of his own cautious intelligence: he is so balanced and sane that he has sometimes balanced his thoughts right out of existence. Useful thinking requires exclusion, emphasis, preferences, commitment, willingness to risk error, because the All is indistinguishable from the Nothing. Though I am far from preferring off-the-wall maniacs to wise and deliberate sages, I feel that a book that has set it sights on capturing in some way the ecological imagination (whatever that may be) will need more than unassailable scholarly acuity and good sense to accomplish its goals. For even unassailability is assailable. *Everything* is assailable if you can't make your sun stand still.

Buell's book is not helped by the fact that four of its chapters, in whole or part, are derived from existing essays and previously delivered papers that have been worked and reworked, while the remainder seemingly have not. The quality of the book's writing ranges from absolutely wretched ("This advantage the analogies of minute realism as grotesque and of eco-centrism as a code of manners underscore in different ways by calling attention to the status of nature-responsiveness as a kind of culture, or rather counterculture, that one must pursue in resistance to the intractable homocentrism in terms of which one's psychological and social worlds are always to some degree mapped" [114]) to decently serviceable (with brilliant flashes), and these disparities have more than an accidental relation to the provenance of the chapters. The more they have been worked over, the better they are. Though there are a few very good chapters, "The Thoreauvian Pilgrimage" among the best (one of the preexisting essays), none would I call "inspired" in the sense that I discuss above. Some of the chapters, such as the one on "place," are almost unreadably dead. This is not because Buell writes the current self-parodying academic jargon (though he uses some of its key terms—and he has no trouble producing his own sesquipedalianisms and tortured syntax) but because the poorest chapters still need more rewriting and revising, or perhaps more rethinking and conviction instead of endless qualification. The book as a whole reads like a loosely thematic collection of essays rather than a treatise on the "environmental imagination," and it more or less just stops when it runs out of gas. The reader is never carried along by an irresistible flow. Of course one does not expect much artistry in a routine scholarly book— one is grateful for mere readability. But this purports to be more than a routine scholarly book—and while it *is* more than routine, it fails as a "book." Finally, despite his thanks to Harvard University Press, they have done him little service: the book is full of typos, including the worst kind: those that form legitimate words that often make some sense (e.g., shippage for slippage); and the copyediting is substandard.

Buell's introductory chapter laments the abstract quality of popular environmentalism and its disconnection from everyday life, and proposes that literary works have the power to make this connection in a primal, emotional way. He notes that American Studies, unfortunately, has paid relatively little attention to environmental writing. He laments recent literary theory's dissociation of the text, the author, and the world, turning them all into ghostly phantasms having little relation to our experience of reality. He would like, in some way, to restore the author as a flesh-and-blood maker of his text, and he goes as far as he can along this line without going so far as to make himself contemptible in the eyes of his more trend-driven colleagues. "Must literature always lead us away from the physical world, never back to it?" (11). This is certainly a worthy question, especially for the chess game of academic literary studies.

The introduction proceeds by examining American writing as a race-gender-class-produced thing that "otherizes" nature but, again, never going as far (to use Buell's words) as "a more radical critic" might go (though it's far enough for me). Charybdis and Scylla are Buell's constant companions: "It is no easy matter to extricate oneself from these biases, to arrive at a more ecocentric state of thinking than western culture now sustains, without falling into other biases like environmental racism" (21).

The chapters on American pastoralism rehearse the familiar duality of the earth as nurturing mother and victim of technological rapacity, a symbol of resistance to culture as well as a model for it. The European colonizing settlers of North America thought admiringly and wrote appreciatively of the New World as a vast unspoiled pastoral wilderness that they nonetheless did not hesitate to despoil by transforming it from nature into culture. But Buell wants to see pastoral ideology as "a bridge, crude but serviceable, from anthropocentric to more specifically ecocentric concerns" (52). He traces the evolution of pastoral ecology from Bartram's travels in the eighteenth century to Edward Abbey's Utah excursions (in *Desert Solitaire*) in the twentieth, with glances at the different point of view to be found in Native American writers. In the course of doing this he provides some keen attention to a number of contemporaries such as Abbey, Leslie Marmon Silko, and Annie Dillard, and their precursors, like Mary Austin. His hope is that pastoral vision—that is, *seeing* the earth—can be transformed from a mode of domination and exploitation to one of green receptivity and ecocentrism.

The recurring uneasiness that Buell feels about literary theory's disjunction between text and world, which he regards as antienvironmental in effect, issues in a desire to restore validity to the ordinary layman's immediate and un-self-conscious response to literary works as representations of reality. The solace derived from nature and nature writing by educated readers and writers such as Wordsworth and Mill is now *infra dig* in academic circles. But Buell does not believe that fictions like Cooper's *Deerslayer* and Faulkner's "Bear" are nothing more than formal or symbolic textual patterns produced in order to be shuffled around in articles and conference papers. What academics would call "naive" responses are still the ones experienced by common readers (a dwindling breed) who read for pleasure rather than professional advancement. (A recent review of a new biography of Steinbeck tells me that *The Grapes of Wrath* continues to sell fifty thousand copies a year. Somebody is apparently still reading "for the plot," but it's not Fredric Jameson.) The discrediting of realism thus exerts a pressure to produce nonecological readings of even avowedly nature-oriented writing. Buell, of course, is quick to add that he "does not deny that they can profitably be so read," but having made this nod toward academic correctness, he manages—as much as he ever does—to hold his somewhat tentative ground in this matter for the rest of the book. Or as he puts it so characteristically, one needs to "reimagine textual representations as having a dual accountability to matter and to discursive mentation" (92). The rest of this chapter (on "Representing the Environment") becomes increasingly casuistical as it tries to sort out the claims of realism and antirealism, fact and fiction. "Without denying that aesthetic realism can validly be characterized from one perspective as a waystation on the path toward total technological control over reality, from another vantage point it signifies precisely the opposite" (113). The returns diminish rapidly.

In a chapter on "The Aesthetics of Relinquishment," Buell looks at the sacrifice of material goods and the sacrifice of the self in the interests of nature, with emphasis on Thoreau, *Robinson Crusoe*, Wendell Berry, Leopold's *A Sand County Almanac,* and others. Here as throughout, there are penetrating *obiter dicta* and flashes of light, but even with all his resources Buell is unable to take wing, remaining earthbound in a way he doesn't intend:

> What distinguishes *Walden* and other epics of voluntary simplicity
> from most traditional narrative plots, including that of *Robinson
> Crusoe,* is that the arrangement of its environmental furniture into

linear corridors through which the protagonist strides becomes less important than what Thoreau suggestively calls deliberateness: the intensely pondered contemplation of characteristic images and events and gestures that take on a magical resonance beyond their normal importance now that the conditions of life have been simplified and the protagonist freed to appreciate how much more matters than what normally seems to matter." (153)

This awkward prose is the downside of an overstocked mind that struggles throughout to rise to a level of truly useful generalization, but because everything is equally true and untrue, the reader is crushed by plausible/implausible, mutually canceling ideas.

With four dedicated chapters and a substantial appendix essay as well as numerous *passim* references, by far the greatest number of pages and the most sustained attention of this book are given to Thoreau, confirming the avowed genesis of the project before it was ill-advisedly pumped up into "the environmental imagination." Buell's expertise on Thoreau is beyond question: with artful deployment of the materials in these chapters (which are generally the best ones in the book), a notable study could very well have been produced. Here, however, it seems to me that there is too much Thoreau for the textual environment into which he has been transplanted from Buell's earlier writings (with frequent additional recircling back to *Walden*). We are told, in the present context of a presumably general argument about environmental writing, a good deal more about Thoreau than many readers will want to know. This excess—through which Thoreau gets qualified almost out of existence—reinforces the pattern already established of too many specificities (which cancel each other out) and insufficient *memorable* generalization.

What we learn is that Thoreau became increasingly sensitized to nature for its own sake as he grew older but retained "the need to organize his observations into aesthetic patterns" (i.e., he was an artist first). His writing reveals a number of ecological "projects" (to use Buell's word), including pastoralism, religio-centric inquest into the relation of the natural to the spiritual, a pursuit of frugality and sustainable agriculture (in the face of the beginnings of agribusiness), and an interest in natural history. In these projects, Buell sees a movement from anthropocentrism toward biocentrism. But in his usual manner, he warns us against too much fetishizing of Thoreau as ecologist. Addressing Thoreau directly, he remarks: "You were groping toward an ecological vision you never grasped; your environmentalism was fitful, your biocentrism half-baked. Fine. We mustn't succumb to mindless hero-worship . . . But neither is it productive to 'demystify' Thoreau and leave it at that" (139).

The chapter that seemed to me perhaps the best in the book was produced from materials Buell had reused and revised a number of times, ending up here as "The Thoreauvian Pilgrimage." We get a solid historical account of the pilgrimages by famous (e.g., John Muir) and not-so-famous people to Concord and Walden Pond and insights into Thoreau's canonization as a culture hero, which was not nearly so much a straight trajectory as a sequence of ups and downs, ins and outs. This is again taken up and developed further in the following chapter, a quite detailed account of the publishing history of Thoreau's works, their editions, sales figures, influence on other writers, and finally the ways in which Thoreau has been made to serve as a motive force in the green ecology of our fin de siècle environmentalism. With lots of factual (as opposed to critical and hermeneutical) information at his disposal, Buell writes more gracefully and has less opportunity to indulge his Hamlet complex. (Since Thoreau's books were actually published in actual years, it is not equally true that they were *not* published in those same years.)

In all these treatments, Buell is at pains to respect the popular mind even as it distorts and exploits larger-than-life cultural icons. In this, he is refreshingly appreciative of the fact that most readers have been "common" readers, and most people are not academic philosophers or literary theorists. (It probably should go without saying that he regards canonizations as both self-serving manipulations and relatively disinterested idealizations.) He therefore stresses a fact that post-deconstructive theoreticians play down or even reject: culture heroes are not just "texts" but actual flesh-and-blood people, and their reputations are as much the result of active mythmaking about them personally as interpretations of their writings. And when Buell explicitly observes "that art is always laboriously produced by real people" (381), I found myself writing "WOW!" in the margin. After years of academic theory spinning, a return to the disparaged vulgar world of "common sense" can seem pretty daring. Of course, Buell never forgets to guard his flanks: "Now, it is hardly clear that restoring a messy intersubjective model of writing and reading will solve all the problems of the world. But we are more likely to make progress if we imagine texts as emanating in the first instance from responsible agents communicating with other responsible agents than if we imagine texts without agency inhabiting discursive force fields" (384). As they used to say in the olden days: "Right on!"

The appendix to this book, "Nature's Genres: Environmental Nonfiction at the Time of Thoreau's Emergence," is a solid study of the sources and influences that helped to produce Thoreau's writings. It is a virtuoso performance by someone whose memory of the exact words of myriad texts is far out of the ordinary. Buell has almost been able to reconstruct the mental library that conduced to the subject position, as we now call it, that Thoreau came to occupy—and he does this without destroying Thoreau's own agency as a person.

But as I warned at the beginning, I have said very little about the contents of this book, contents that are full to the point of overflowing. By the time I reached the final page, however, I had forgotten almost all of it. What remained sharply delineated in retrospect was the character of the performance itself. To me, it's a troubling phenomenon, an exemplar of "the postmodern condition." In his book of that title, Lyotard told us we are living in an information age in which data is power. But Buell's nervous and fretful databank is a symptom of powerlessness made all the more vulnerable by today's nonjudgmental egalitarianism, its equal-opportunity dialectic. One senses a prophetic yearning and a desire to hold opinions, which now and then manage to ooze through the crevices of the "beautiful mosaic" (as New York City mayor Dinkins once optimistically described "diversity"), but the dialectical data keep multiplying like a computer virus, stanching the flow of conviction—which has already been weakened by a fear of academic censure. The "person" behind the book comes off as a socially constructed sorcerer's apprentice who has unleashed a power he is ultimately unable to control—and he gets trampled in the process. I say without irony that I hoped he would prevail.

CHAPTER NINE

The "Environment" Is Us

B OOKS DEALING WITH ECOLOGY and environment are now a vast industry, an avalanche of information and opinion that exceeds anybody's ken. The "environment" itself keeps growing, enlarging, encom-
passing, so that the environment of 1998 is a very different thing from
what it was on the first Earth Day in 1970. The sheer number of disciplines that have evolved since Aldo Leopold's landmark *A Sand County
Almanac* of 1949 is startling—environmental medicine, environmental
history, environmental engineering, environmental ethics, social ecology,
green travel, green farming, conservation biology, eco-feminism, ecocriticism, animal rights, to name a few—exceeding in subtlety and complexity such early concerns as emissions, toxic waste, acid rain, and cancer
clusters. On the World Wide Web alone the information is daunting,
hopeless, beyond belief.

In fact, the term "environment" now seems inadequate, a misrepresen-
tation of the current state of affairs. After the Industrial Revolution, human
beings came to be seen as more or less autonomous creatures who had
been placed in an "environment" that they could use as they wished or
even, in some perverse sense, do without. Understood rather literally, the
environment was the stuff that surrounds us: factories, automobiles, trees,
skies. Now, however, the center around which the environment wraps is
getting smaller and smaller (or larger and larger) as what formerly seemed
adventitious to the Imperial Self begins to look more and more essential
to its very constitution. The "environment," as we now apprehend it,
runs right through us in endless waves, and if we were to watch ourselves
via some ideal microscopic time-lapse video, we would see water, air,
food, microbes, toxins entering our bodies as we shed, excrete, and ex-
hale our processed materials back out. Western through and through, I
say this without any flirtatious gestures toward Zen, any practical sense

that individual things are an illusion (in a philosophic sense, *everything* is an illusion), or any lapse of faith in the Imperial Self. The "ecocentric" rant ("I'd sooner shoot a man than a grizzly") that briefly served Edward Abbey and Dave Foreman with such bravura showmanship has had its day, and now Abbey and Foreman seem as imperially selved as anybody else, if not more so. (Though Foreman has since become a mainstream eco-pussycat.)

Three almost randomly chosen new books from the environmental deluge work together, when read as a group, to heighten one's sense that the environment has ceased to be a wrap and looks more and more to be the very substance of human existence in the world. The first of these takes an "inside" approach (via subjectivity), the second an "outside" approach (via public policy reform), and the last is a philosophic overview of the influential theory of "deep ecology." The writing styles and *mentalités* are as unlike as their contents.

Bodies in Protest: Environmental Illness and the Struggle over Medical Knowledge[1] by Steve Kroll-Smith and H. Hugh Floyd is a study that comes perilously close to disaster but somehow manages to add up to more than its liabilities. Written by two professors of social science, the book is weighed down by portentous Foucauldian melodrama, not merely employing but repeating to distraction every cliché in the cultural studies lexicon, so that it often reads like a boilerplate whose blanks have been filled in with its ostensible subject. The authors seem to regard the clichés as the essential part of their achievement, but if they had all been left out, little would have been lost—and the gain might have been a shorter, more impressively "original" essay. Both their cultural studies theme (endlessly repeated) and their somewhat lumbering social science style are typically represented by the following:

> Throughout this book, the idea of EI [Environmental Illness] as a new way of knowing the body in its relationship to built environments is revealed in the activities of ordinary people who claim the right to theorize their bodies and thus shift the social location of theory construction from experts to nonexperts. The contours of this new knowledge become more visible as we record how these theorists change the definitional strategies of science from a focus on nature and the person to a critique of society. Finally, the political efficacy of

MCS [Multiple Chemical Sensitivity] is measured by its rhetorical power to convince the world that modern bodies and the environments they build are undergoing profound change. (66)

Or as they put it elsewhere: "Environmental Illness is a story constructed by nonexperts about human bodies in somatic dissent against a material world saturated with commodities promising to make life easier and healthier [but doing the opposite for many people], and the body itself more attractive" (136). This is surely academic newspeak (now sounding somewhat shopworn) "constructed" according to the most trendy canons of academic power. Behind the locutions are observations worth heeding, but the language itself compromises much of the insight. To take just one example: the very title of the book, *Bodies in Protest,* is an imprecision dictated by current academic locutions. Although the authors say with some plausibility that environmental illnesses have produced a distinctive narrative rhetorical style in the people who suffer from them, in no sense of the word is it the "bodies" that are "protesting" but the people (subjects?) who inhabit these bodies. The real "rhetoric" of bodies is burps and farts, bleeding and drools, not linguistic narrative techniques—which emanate from the consciousness of "people" (now an obsolete and retrograde word in a Foucauldian world of "discourse formations" that produce the mouthpieces that speak them).

The authors' basic claims, however, are convincing: as a result of a vast array of modern materials, from carpets to treated wood to ventilation systems to perfume, a tremendous web of chemical sensitivities has been produced that exceeds the available verification methods of contemporary medical science. The people who suffer from these maladies—a substantial number, it should be added—are more often than not told by their doctors that tests reveal nothing wrong with them and that it may indeed be all in their heads. But the large number of stories and quotations that the authors supply from their interviews with chemically sensitive people almost never give the impression these ailments are principally psychosomatic. The intelligence and rhetorical skill with which the sufferers describe their conditions are powerfully persuasive and constitute what the authors refer to as an "alternate rationality." Once they have been deserted or insulted by their physicians, these people research their own conditions and, without abandoning mainstream medicine, use the techniques and data of the sciences to present an alternate analysis that is nonetheless far removed from New Age moonshine. The authors' analysis, however, is predictable: "Medicine works closely with the state to define and regulate bodies in the interest of cultural and capital production" (32), and they substantiate this cliché with a reference to Foucault. It's not as though such a concept were altogether useless—but it's not the reductive, totalizing, gospel truth they imply by their interminable metaphoric repetitions (e.g., "Biomedicine is charged by the state with writing the somatic text" [48]).

Kroll-Smith and Floyd seem uncertain whether MCS and EI are a linguistic thing, a psychological thing, a political thing, a bodily thing, a moral thing. They make passing remarks to the effect that EI "joins a mind to a body that is no longer readily intelligible by cobbling together clusters of words to tell a story of disease," even though they don't really seem to think that these stories are just "stories." But narratology is another academic fetish right now, and the authors are trapped in conflicting paradigms that undercut the substantial message they have to convey. On the one hand, they admire these sufferers for standing up to a colonizing medical profession, but on the other hand they are a little jittery about committing themselves above and beyond the licenses of their boilerplate. They come off as most at ease when they can sing professionally accredited arias about power, the state, bodies, etc., etc.—in other words, when they can sound like everybody else.

Given these vacillations, their conclusions seem admirable. If mainstream medicine won't legitimize environmental ailments, the sufferers themselves must learn how to do so. Not only must they be able to explain themselves effectively (which many already do), but they must engage the sympathies of their listeners and the social institutions of which they are a part. The authors provide graphic evidence of changes made in the workplace and elsewhere as a result of the willingness of coworkers and friends to acknowledge the realities of kinds of suffering they themselves do not experience. Cases of environmental illness are more and more frequently being won in courts as a result of *legal* rather than *medical* criteria. "Ordinary people are fashioning a new form of rationality to account for changes in their bodies, blurring the boundaries between layperson and expert" (197). Although "a new form of rationality" seems like another professorial extravagance, Kroll-Smith and Floyd have some-how managed to avoid being totally steamrolled by the heavy-duty "con-struction" equipment of their profession.

Thinking Ecologically: The Next Generation of Environmental Policy,[2] edited by Marian Chertow and Daniel Esty, is far away in style and sub-stance from the previous book. The essays presented derive from an on-going project at the Yale Center for Environmental Law and Policy. All are judicious, reasonable, and clearly expressed in the well-tempered lan-guage of policy studies, the subjects and points of view bringing to mind the annual *State of the World* volumes from the Worldwatch Institute, al-though the latter tend to be more impassioned and programmatic. The world being put under the microscope here is not the "subjective" one of individual experience but the "objective" public world of politics, indus-try, lifestyles, employment, and international relations. The contributors discuss issues such as industrial ecology, ecosystem management, prop-erty rights, land use, technology, ecological law, automobiles, energy prices, and the global economy. There is a wealth of information about what is going on in every imaginable area of practical ecology, and if the book can be said to have an ongoing theme or viewpoint (which I think it does), the message has to do with interconnectedness and cooperation. As they cogently demonstrate, the era of governmental fiats from on high and piecemeal solutions to environmental problems is now winding down in favor of more integrated policies whose aim is to incorporate environ-mental considerations and costs into the products, processes, and services that drive the global economy, considerations that are gradually being naturalized in a sort of corporate superego. In one of the most interest-ing of the essays, "Coexisting with the Car," Emil Frankel outlines the failure of government regulations to solve the problems caused by the rapid growth in private auto ownership.

> Policymakers must recognize reality: Americans cannot be forced out
> of automobiles by regulation nor cajoled into using them differently
> by highminded appeals to sacrifice personal convenience for the good
> of the whole. But at the same time, we cannot drive away and forget
> the problem. The key will be making car and truck travel pay for
> itself. When the full costs of pollution, congestion, and habitat
> destruction are factored into driving, incentives for change—in per-

sonal behavior, corporate transport decisions, and technologies—will be created. (191)

In sum, the contributors see "environmentalism" not as an external pressure but as an increasing strand in the fabric of every societal activity.

Primitives in the Wilderness: Deep Ecology and the Missing Human Subject,[3] by Peter C. van Wyck, is the most inclusive and intellectually sophisticated of these three books—inclusive because it subsumes the private and public foci of the first two studies and sophisticated because its perspective is essentially philosophic and self-reflexive. Van Wyck, as humanist, has mastered (for good or ill) the language of cultural studies that Kroll-Smith and Floyd bumbled so heavy-handedly and uses it as the medium for analysis of the crippling deficiencies of deep ecology as a type of environmentalism. Van Wyck's prose, however, is far from exemplary, blighted by numerous obscure passages (endemic to cultural studies), occasional solecisms and syntactical blunders, and deficiencies of copyediting. Still, if you can tolerate the lingo, his is an impressive critique. Van Wyck, like Murray Bookchin and others (including me), recognizes deep ecology for what it is: a univocal, absolutist, messianic, misanthropic, pseudo-primitive rejection of contemporary life. Deep ecology

> lifts and relocates a contested and confused modern subject from its structured relations to ideology, politics, the unconscious, and so on, to a smooth, noncontradictory ecological space.
>
> No longer a potential site of resistance, the ecological subject is undifferentiated from its context. This subject is no subject at all; it becomes a desubjectified organ of Nature. It is a dream of a post-historical subject and its pathology is that of a transcendental narcissism. (105–6)

The jargon, admittedly, is relentless but usually not impenetrable. The roots of deep ecology can be found in Aldo Leopold's "land ethic," which (to quote from *A Sand County Almanac*) "changes the role of *Homo sapiens* from conqueror of the land-community to plain member and citizen of it." As developed in recent years, this has come to mean that human beings are just another biological element in the ecosystem and that to privilege them is to be "anthropocentric." Hence, Foreman and Abbey's claim (in differing words) that they would just as soon shoot a man as a grizzly. Deep ecologists speak of the "intrinsic value" of everything rather than the "instrumental value" we place upon things we suppose to exist for our own benefit—but the list of implied exceptions is rather large, including cancer cells, the AIDS virus, bubonic plague, and cockroaches. And since trees are unable to defend their own "intrinsic value," obviously they need spokespersons trained to "speak for the other." In a Foucauldian ethos in which all speech is a power ploy, somehow the deep ecologists fail to notice that speaking for others is like speaking for God—the biggest power ploy known to humankind. These bourgeois couch messiahs sing the praises of hunter-gathering (farming was the beginning of exploitation), fantasy-driven matriarchal cultures, goddess-worshiping cults, indigenous peoples' primitive harmonies with the land, and other golden age fatuities.

It is van Wyck's aim to show that deep ecology abstracts human beings from the political, social, and economic flux in order to position them in a fixed and timeless "nature" arbitrarily defined from infinite possibilities. That there may be a hidden agenda in such choices (dictated by time, place, culture, parents, psychological history, education) does not seem to occur to these ecological thinkers. The human subject, i.e., the person (everyone besides the ecologist), simply disappears as an individual consciousness and becomes an anonymous member of one species among many. The deep ecologist fails to notice that, far from being ecocentric or biocentric, he is as anthropocentric as anybody else, since any system of thought is a strictly human production determined by societal and personal contingencies. Furthermore, no species is ecocentric, because survival depends on looking out for one's own interests. Do birds suffer angst as they shit upon your head?

More traditional than van Wyck's language is a quotation he gives from George Bradford: "'Ecology as science speculates, often with profound insight, about nature's movement and the impact of human activities on it. But it is ambiguous, or silent, about the social context that generates those activities and how it might change. In and of itself, ecology offers no social critique, so where critique flows directly from ecological discourse, subsuming the complexities of the social into a picture of undifferentiated humanity as a species, it goes astray'" (61). In our multicultural world, the ecological situation differs drastically not only from country to country but from cultural persona to cultural persona. The "we" of deep ecology, as van Wyck likes to point out, is far from a unitary one. The end of the line for "we" speaking for the "Other" and for unilateral definitions of reality is, of course, the manifesto of the Unabomber. But we've seen plenty of other messianic precursors who dynamite research laboratories, free animals from enclosures, blow up the World Trade Center and the Oklahoma City Federal Building, shoot abortion doctors, and so forth. These saviors are more than anthropocentric—they're basket-case narcissists for whom there is no "Other" at all.

Van Wyck looks at radical ecology's suspicion of reason, the Enlightenment, and science. He remarks with irony, "Peel back the layers upon layers of history, technology, culture, modernity, and on the inside, the very center, there can be found the kernel of the real human: the ecological subject" (104). But, he asks, who is this ecological subject? Alas, nobody answers the call. "To say that organisms (humans and others) are produced discursively [that is, through adaptation and the institutions of culture], that they do not preexist themselves, is to radically change what can count as 'nature'" (119). Although Murray Bookchin's social ecology—on a track similar to van Wyck's—avoids many of the pitfalls of regressive and phony primitivism (and Bookchin hates deep ecology with a passion), his own suspicions of centralized technology and hierarchical relations between institutions and people entail other unrealistic limitations to a solution of the problem that used to be called "man and nature."

Van Wyck tries to resolve some of this problem's antinomies by concluding his book with an account of "situated knowledge," particularly as developed by Donna Haraway. The goal is a limited objectivity that does not make eternal truth claims but that also does not disparage the validity of subjectivity, suppressing instead invidious sets of alternatives like anthropocentric and ecocentric, self and Other, subject and object. "The claim that Haraway's objectivity makes is not one of detached truth-seeking from some imaginary point above the fray, but of limited, localized, and embodied knowledge," a kind of "'conversation' bounded by affinity and complicity" (123), a living amidst contradictions.

Finally, van Wyck borrows from the philosopher Gianni Vattimo the notion of "weak thought": "The weakening of thought in this sense refers to the weakening of the 'autonomous pretensions of reason' [quoted from Iain Chambers]. This 'weak' turn does not imply the absolute death of reason, or the end of value. Rather, its claim is on an always reflexive position with respect to reason and value. It is a kind of thought and practice that must always remain aware of its own artifice, its own locality, and its always limited scope" (129). Admittedly, this vacillating "solution" is not as satisfying as tablets from on high—but the contemporary problem of belief is itself the problem of the unbelievability of truths from on high. So this may be the best we can do for now, short of Unabombers and Timothy McVeigh.

All three of these interesting books share a sense of the gradual disappearance of the environment as an "out there" thing. As the subjective experience of the chemically ill is bit by bit objectivized, as corporate policy internalizes the environment into an essential ingredient of production, as the supposed objectivities of deep ecology are seen more plausibly as negotiations between selves and Others, the conflict between people and the environment looks to be moving toward an awareness that "the environment," after all, is really us.

CHAPTER TEN

Ecology and Ecstasy on Interstate 80

Loth to believe what we so grieved to hear.
For still we had hopes that pointed to the clouds,
We questioned him again, and yet again;
But every word that from the peasant's lips
Came in reply, translated by our feelings,
Ended in this,—*that we had crossed the Alps.*

WORDSWORTH, *The Prelude,* VI, 586

O N MARCH 28, 1996, I packed up the car in preparation for a five-thousand mile automobile trip to the Southwest and California that would take me away from home for at least three weeks. The plan was to visit Tucson, Los Angeles, Davis, and Reno to see a number of friends and family members as well as to explore a few potential warm spots to which I might move in order to escape once and for all the harshness of Chicagoland's winters. The drive would doubtless be bittersweet, a lonely solo unavoidably retraversing many places that my now dead wife, Gloria, and I had visited together—in our unflagging happiness—sometime around 1985. The final stop would be Reno, where a little book-signing party was to be sponsored by the University of Nevada's English department to celebrate the publication of *The Ecocriticism Reader,* the fruit of half a dozen years of rewarding editorial collaboration with Cheryll Glotfelty.

I packed up my still quite new, rather spiffy, Saturn SL-2 with more than enough provisions and equipment for meals in motel rooms, including a hot plate, cans of wholesome soups and chilis, coffee, oatmeal, low-fat cookies (and other necessities of a health nut's diet), utensils, gifts, clothes for various climates, and twenty-four compact disks loaded into two twelve-disk magazines that could be inserted into the compact disk changer that I considered an essential option in the car's purchase. Some long and lonely days inevitably lay ahead.

The next morning I took off for Springfield, Missouri; then Amarillo, Texas; Las Cruces, New Mexico; and on the fourth day—after passing a moonscapish formation of rocks and boulders in southeastern Arizona—I eyed with slight nervousness the faint nimbus of orange sky that greeted my arrival at Tucson. My obsession with ecology had begun in the early seventies when Gloria and I (deceived by the benign direction of summer winds during house hunting) were ignorant enough to have bought a little farm in northwest Indiana, captivatingly beautiful but also the most polluted place I have ever dwelled in my life. In our dispiriting and futile effort to contend with the physical and mental symptoms caused by apocalyptic quantities of effluents from the steel mills of Gary, we quickly turned into hypersensitized canaries, acquiring a new awareness of even the most miniscule amounts of insalubrious air. Now, all my travels resembled the food tours of decadent gourmets, although in my case the ingestions derived not from eager tastings of the sophisticated concoctions of four-star restaurants but from involuntary inbreathings of complex toxic bouquets, the particular particulates and waste products of distinctive industrial outputs in cities and countrysides all over North America and Europe.

Tucson's air turned out to be still relatively salutary, although a far cry from its celebrated purity of yesteryear. But it was better than Chicago's and infinitely better than anything breathable around Los Angeles, so I felt Tucson offered a real possibility for relocation. After a few rewarding days exploring landscape and housing, I headed northwest on the I-10 to Phoenix, which proved a very different kettle of fish. Halfway there, though nothing obvious was to be seen around me, I began to experience a tightening up of the sinuses and throat—what people call "flu-like" symptoms—as well as the familiar signs of a pollution headache. (But in my years of experience the most toxic air pollution has tended to be partially or totally invisible.) Once into Phoenix, however, I began to feel well again under sunny blue skies. I now could see that the northeast wind was blowing a vast dark orange plume of smog to the south and west of the city, a plume whose outer edges I had probably briefly traversed en route from Tucson. This plume, broadening rapidly into a wider and wider triangle as it expanded from its source, accompanied me all the way to Blythe at the California border, a distance of roughly 175 miles. As it gradually dissipated into the California desert, I could begin to see, leaving Blythe, what I took to be the outer edges of pollution from Los Angeles, still about 225 miles to the west, extending welcoming arms to embrace me in a chokehold. The smog became increasingly intense as I got to Palm Springs, a polluted desert oasis, and as I approached Riverside, one of the most notoriously smoggy areas in the United States, the entire valley from Los Angeles eastward appeared enveloped in a cloud of toxins.

For a few days I settled with friends in Fullerton, near Disneyland, the same friends who had told me on each previous visit that in Fullerton they were "not bothered by smog," an ambiguous report I never could fathom. Did it mean that Fullerton itself was exempt from smog or that my friends recognized its presence but were never personally "bothered" by it? Whatever they intended, my days there were marked with virtually nonstop headaches and malaise, the skies were orange, and one of the friends who claimed not to be bothered fell into drowsy states several times a day that segued into brief catnaps. It struck me as more than a funny coincidence that these naps corresponded perfectly with my worst headaches and "flu-like" symptoms. And indeed, I myself had several bouts of pretty irresistible drowsiness during the week I spent in the greater Los Angeles area, even after some exceptionally good nights of sleep.

When my visit had ended, I headed north on Interstate 5, the Santa Ana Freeway in Los Angeles, which soon crosses the San Gabriel mountains and makes its way up the San Joaquin Valley, the vegetable-growing capital of North America. The wind was from the south and the plume from Los Angeles was dispersed into a blurring haze throughout the valley, almost as far north as the imaginary line one could draw from Salinas to Fresno, about 225 miles from L.A. I vividly recalled stopping for gas on an earlier trip up this route with Gloria and my Fullerton friends, issuing my customary complaints about the shockingly bad air, complaints which often rub people the wrong way, impatient with what strikes them as sheer crackpotism—since *they* claim not to be "bothered." Here, perhaps seventy-five miles north of Los Angeles, the gas station attendant had completely surprised me by volunteering the information to my party that this polluted mess in the middle of nowhere was Los Angeles smog! I regarded him as a secret ally.

As I reached the Sacramento/Davis area, about ninety miles east of San Francisco, the skies looked good, and I felt okay. The winds were carrying Sacramento smog far to the north, beyond my projected route, a sharp contrast with my experience several years earlier when I flew to Sacramento from Chicago and was surprised to find the ground completely obliterated by orange smog as the plane circled in for a landing—a more representative picture, as I since have learned, of what happens to smaller cities as they grow into large ones.

So there it was. I had already covered about three thousand miles, and as I traversed the great open spaces of our heroic pioneering West, everywhere I looked were miles and miles of toxic air, the fruits of expansionism and technology, fruits that, in my mind, were making millions of people feel wretched every day (without their knowing why) and contributing to long-term, often fatal, diseases which one day would suddenly appear as if from nowhere to do them in. Electric power plants, oil refineries, steel mills, millions of automobiles, dry-cleaning plants, suburban lawnmowers, Jet Skis, snowmobiles, sports-utility vehicles, copper smelters—you name it. Meanwhile, trees that produce oxygen were being cut down to produce Big Macs, methane gas from cattle, and pollution from animal wastes; water was being fouled by paper mills and oil spills; fish killed by pesticide runoff—you know the story: the nightmare of technology, the inheritance of the Industrial Revolution. "Abundance makes me poor," as one of Ovid's wiseacres would have it. To suppose technology is not among contemporary society's chief critical problems would be to live in a fool's paradise, to be permanently out to lunch.

After spending two days with my friends in Davis, I was concerned about leaving early enough to arrive in Reno before dark, perhaps a two-hour drive now that the speed limits have been raised to 70 or 75 mph. I said goodbye, hopped into the Saturn, turned on the compact disk player—my salvation—and sped off into the not-yet sunset. Davis and Sacramento are flat, flat, flat, but the Sierras' foothills were not far to seek, and within an hour I could see the road starting to climb. I was already feeling a little inebriate, having just heard Beethoven's *Fantasy for Piano, Chorus, and Orchestra,* that curious mélange of styles and themes that eventuated in the "Ode to Joy," rapturous when done by the right conductor but a flop in less capable hands.

My spirit soared as the road reached higher and the air thinned out, the rich vegetation of the western side of the Sierras took on the intense green of a late afternoon in mid-April, and the curves got sharper. Like Dr. Johnson in his primitive horse-drawn coach en route from London to

Edinburgh, I felt there was nothing more exciting than racing along a highway at top speed. Sweeping through woods edged by sudden declivities with panoramic mountain views that a solo driver dare not examine too minutely, I was a little surprised to hear the opening orchestral chords and shouted "kyries" of Beethoven's *Missa Solemnis,* which for the moment I forgot I had loaded into the CD player's magazine. This was the great performance by Bernstein and the New York Philharmonic, recorded in 1960, certainly one of the best on records, issued first on LPs and now (along with the aforementioned *Fantasy*) digitally remastered on Sony compact disks, more clearly audible than ever before. The *Missa Solemnis,* one of the dozen or so most sublime musical creations of Western culture, sounding even today grotesquely cacophonous in its Dionysiac syncopated frenzy, requires—like the symphonies of Bruckner—the most skillful of conductors to hold it together, or else it can simply fall apart into a series of disconnected and stumbling episodes. Caught in its mania, I was driving faster and faster, struggling to negotiate the curves and forced over and over again to slam on the brakes to avoid going off the edge. Monumental vistas were unfolding, my excitement level kept rising, all my senses were being stimulated at once.

I was reaching seven thousand feet, I had just passed the stunning sights of Emigrant's Gap—when the most stupendous moments of the entire *Missa* commenced—the high-speed fugue on "et vitam venturi" in the Credo, insanely executed by Bernstein and his chorus about twice as fast as the speed of my car, music that at mere ground level always left me in a heap, a pulp, a burned-out shell of a person. Now, at high speed and high altitude, I reached a pitch of excitement almost hysterical. I was traversing Donner Pass, skirting Donner Lake, trying to take in this incredible panorama of sights from the world and sounds from the car, thinking inevitably of the tragically fated Donner party, half of whom perished right there over a hundred years ago when California was little more than a string of Spanish missions founded yet another hundred years earlier by Father Junipero Serra. The strain (to borrow from Donne) on that "subtile knot" that joins spirit and flesh reached the breaking point as the fugue drew to its frenetic close, and I burst into tears of sensory overload while pressing down on the brake pedal to keep my car from swerving off the road. The slant of light from the late afternoon sun was still painterly, coloring the multilayered geological cuts through which the roadway passed and sharpening the relief of trees against mountainsides, of road curves against mountain passes.

It was at that precise ecstatic moment that I experienced my ecological epiphany. Though it didn't come from God, John Muir's Sierran hosannas to what he called divinity and what Shelley more aptly called "the intense inane" were never far from my mind. It derived, rather transparently, from easily identifiable components of some of the most major intuitions and experiences of my reflective life, but now, like the treescape's sun-illuminated relief, my ordinary horizontal thinking had been shot through by a vertical bolt of lightning-insight, casting the mundane into the sublime by making it possible to think a host of thoughts simultaneously.

And exactly what did the mountains have to say?

Put somewhat baldly: "Everything human is technology!" Perhaps one would prefer to say not *everything,* but almost everything. The spiffy car I was driving, with its air bags and antilock brakes. The compact disk player, the disk with the *Missa,* the analog tapes from 1960, the recording process itself, the original performance with all of the manufactured instruments and trained voices, the system of producing and distributing the disks, Bernstein's jet-setting existence and materialities, Beethoven's own life, his music paper, his pens, his piano, the musical "logic" that enables composition. As for his deafness, more advanced technology might have alleviated it, changing everything. Then there was the roadway on which I was driving, the incredible engineering feats that cut through the mountain passes, and the geological layers thereby revealed (illuminatingly examined by John McPhee in *Assembling California*). The mountain views were themselves the fruits of technology—of decisions about where to put the road and the angles of vision that resulted, of the appearance of the layered roadcuts and their contribution to the esthetic experience. Mostly everything about me and my life had a technological connection as well: the clothes I wore, the computer I used every day, the manufactured food I ate, my shaving equipment, electric toothbrush, wristwatch, the crowns in my teeth, glasses, orthopedic shoes. Maybe Donna Haraway was right: we're already cyborgs, half organism, half prostheses; half nature, half technology. Surely the uplift I felt at this landscape required a healthy body, good food, bourgeois nurturing and education, modern equipment and appliances—all from technologies. Indeed, that I had survived childhood to become a physically fit adult after several potentially fatal diseases was due in large measure to the biotechnologies of medicine.

From the first stone tools of paleolithic peoples to the latest modem access to the World Wide Web, from the poisoned Roman populace who drank water in leaded cups to the irradiated corpses from Chernobyl—the good and the bad of human life were mostly technology. Technophobes may praise the Amish for their simplicity, but what distinguishes the Amish derives not from eschewing technology but from fixation upon one of its earlier stages. Why should any particular phase of technology—or of evolution, for that matter—be thought of as more "natural" than any other? Are animals bred by humans to pull wagons more "natural" than machines designed to do the same thing? Without technology—the payoff from opposable thumbs—human beings would never have been able to lift themselves out of primal animal existence. Even the most nature-committed postmodern adventurers are completely dependent on the latest inventions. Edward Abbey, for all his chest-thumping bravado in *Desert Solitaire,* was not as solitarily self-creating as he liked to make out. Floating down the Colorado River in his inflatable raft stocked with tinned and dehydrated foods or roughing it in Havasu with telephone-ordered provisions mailed from the grocery store at the Grand Canyon, he was as much a child of technology as the bourgeois tourists he satirized in the recollections of his ranger days at Arches National Park. Today's wall climbers and backpackers would hardly exist without L.L. Bean, Gore-Tex, Rockport, water purifiers, camping stoves—and their four-wheel-drive gas guzzlers.

The sciences of ecology are themselves enabled by devices to measure pollutants in air and water, pesticides in vegetables, radiation from failed power plants. The air pollution in New York, Chicago, northwest Indiana, Phoenix, and Los Angeles may have tainted my life with an ongoing malaise, but my epiphany on I-80 made it plain that the bad and good were so inextricably tied together that to be against technology was to be against human life itself. I thought of the absurdity of Max Oelschlaeger's nostalgia for hunter-gatherers in *The Idea of Wilderness* as the decadence of a technologically pampered bourgeois philosophy professor. The war of the well-feds against technology looked less like Ludditism than like the religious and political cults of Jonestown, Waco, and Oklahoma City.

Indeed, the "ecocentrism" and "biocentrism" of the deep ecologists has an alarming resemblance to the right-wing power ploys of misanthropes like Phyllis Schlafly and Pat Robertson. If the reactionary right can be said to fear and hate adult consciousness and to love only what they can safely ventriloquize and control without backtalk—namely God and fetuses, creatures that express the fantasies of their ventriloquist-creators—then the deep-ecological left can be said to constitute their mirror image, with the Unabomber their basket-case *doppelgänger*. If the religious right spouts self-regarding, repressive, and maudlin essentialisms about the will of God, about the *real* nature of men and women, sex, marriage, and family life, the deep-ecological left essentializes transient stages of evolution and speaks of ecosystems, natural habitats, wilderness, animals, and "nature" as though they were platonic ideas, fixed for all time instead of evolving aspects of a universe without stasis, an evolution no less "natural" after the Industrial Revolution than before.

In an evolutionary universe, things adapt or perish, so nature is anything survivable, not just the familiar species that happen to have populated recent centuries or our own more recent childhoods to provide deep ecologists with "eternal" platonic forms. Instead of mendacities about "the will of God" and human normalities, the deep ecologists have their own mendacities about "speaking for the Other," for trees and wildlife and mountains, just one more disingenuous stratagem of the will to power. Their counterpart to God's will is the notion of "intrinsic value," which replaces the narcissistic humility of religious extremists with denunciations of anthropocentrism for its "instrumentalism," a relationship to the natural world—it is claimed—that fails to recognize the intrinsic value of other species. "Intrinsic value," however, is itself an instrumental oxymoron. Its purpose is to foreclose conversation, like references to God, and to establish the "innocence" (i.e., reverence for life) of biocentrists vis-à-vis the selfish predatoriness of anthropocentrists.

But nobody is innocent. *To be alive is to be a murderer!* Recycling is the master algorithm of the universe. The only authentic biocentric act is suicide, freeing up finite matter for the benefit of others. Everything is instrumental except survival of oneself. Inasmuch as all value is conferred by a reflective consciousness, nothing has *intrinsic* value except a reflective consciousness reflecting upon its own incarnation. When Dave Foreman tells us that grizzlies are more "important" than people, or when Phyllis Schlafly tells us that atomic bombs are the gifts of a wise God (to keep down non-Western, non-WASP adults so North American fetuses can be turned into religious conservatives), we learn nothing about either intrinsic values or God's will, only the bad news we already knew: that Foreman and Schlafly are misanthropic powermongers for whom "nature" and "God" are not-so-secret agents of desire.

The master motive for human beings was always human survival and its attendant human interests. Aldo Leopold's land ethic and his role as a father figure of biocentrism were necessary and heroic developments after the destructive technologies of World War II, when a new awareness of suicidal human depredation (the result of shortsightedness and ignorance about what makes for survival) was desperately needed. But fifty years later, Leopold's biocentric commandments have become pious clichés for undergraduate term papers and for political reappropriation by the bourgeois anti-bourgeois left. Whereas Leopold could speak pioneeringly in the war-shattered forties about "reappraising things unnatural, tame, and confined in terms of things natural, wild, and free," in the late nineties these words function merely as invidious political terms, however benign their original intent—or as poetry. As ontological concepts, "natural," "wild," and "free" now seem almost meaningless, if not preposterous. Today, biocentrism and "intrinsic value" can be seen simply as other forms of human interests, as "anthropocentric" as the rest, no more disinterested than acts performed "for the greater glory of God." Even in the case of Leopold, biocentrism was offered as an enhancement of human life (both practical and esthetic), since human life depends on a particular ecological mix that war and unbridled capitalist/communist technology have threatened to destroy. From a human perspective (and what other do we have?), the wilderness (a recent invention) and grizzlies (a recent obsession) aren't being preserved "for their own sake," but because certain people like them, need them, or regard them as necessary for a better sort of human life. If "existing for their own sake" were the real criterion of "intrinsic value," then cockroaches and cancer would be as entitled to exist as anything else. If wolves can be reintroduced into Yellowstone ("for the ecosystem"), why not smallpox into the Western world?

As I moved at high speed through the wondrous Sierras while the sun declined in the late-afternoon sky, my electrifying sense of the primacy of technology—ever in need of control—and the ineluctability of anthropocentrism—which does not always recognize its own survival interests—

was intensified by recollection of the passage from *The Prelude* that I have quoted above. What so shocked Wordsworth and his party, psyched up as they were by the notion of traversing mountains that had seemed dauntingly majestic from a distance, was to learn that they had already crossed the Alps! In Wordsworth such a realization inevitably leads to a passage extolling the wonder and power of the human imagination, a faculty that half creates what it beholds (and, one might add, that makes nature in its own image, just as it has always done with God). But I thought too of John Muir's seemingly ecocentric hosannas in these very Sierras: "every crystal, every flower a window opening into heaven, a mirror reflecting the Creator," a passage (one among many) so different from Wordsworth's yet finally as anthropocentric as his. Wordsworth's was an unabashed celebration of human faculties even greater than the Alps; Muir's, an erotic ecstasy that figured itself in the orgasmic language of hallelujahs—for nothing is quite so human-centered as imagining a universe made for our delight by a deity that has given his all.

My own epiphany was like Wordsworth's, only more so. These wondrously beautiful "Sierras" (which here can stand in for all of "nature") had no real existence of their own. There may indeed be Sierras underlying *my* "Sierras," but as philosophers from Thales to Rorty have made apparent over two millennia, we know very little about them, and most of what we do know comes from the natural sciences. We live in a world of perceptions and appearances and, for us, appearance is reality. When I read his great book, *My First Summer in the Sierra*, I could not help picturing Muir amidst the mountains as a dust mite stumbling along inside a rich piece of velvet. Caught deep within the individual strands of pile that make velvet look and feel smooth to a comparatively gigantic human observer, the dust mite doubtless sings hosannas to the grandeur and mystery of each rough and monumental peak, clearly the product of some sublime and powerful deity. When human beings leave their "scenic vista" highway lookouts, miles away from velvety mountainside forests, and wander closer in to inspect the rugged mountain trees from inside, are they not like dust mites in velvet, scrounging around in the undergrowth of giant conifers? And when human beings behold the Sierras from a jumbo jet at forty-two thousand feet, are the dune-like undulations seen from above any the less "real" than the majestic peaks seen from below? And what do astronauts think, hundreds of miles above the earth, the Sierras now flattened into a splotch of color? Why sing paeans to majestic Sierras at all, any more than to strands of velvet pile, if the majestic Sierran peaks are no more "real" than (or just as "real" as) the flattened blotches from outer space? Velvet pile is as awesome as Sierras if you happen to be a dust mite. Sierras are as innocuous as velvet pile if you happen to be an astronaut. Why privilege the ground-level perceptions of a particular species (i.e., us) as representations of the *real* reality? We squash out creations as sensational as Sierras with every step we take: not being small enough to appreciate their microscopic majesty, we fail to sing songs of reverence to mitochondria or particles of clay. Selective attributions of deity are the very essence of anthropocentrism. Assessed from the totality of possible viewpoints, everything would be equally sublime or mundane. Is the world really a spectacular show designed for humanfolk, staged by a cosmic Ted Turner, biggest cable operator in the universe, who stays awake at night worrying about the fall of sparrows and other violence and sex aired daily on his infinite channels?

"Nature," like everything else a contested site (as they now say), is the technological production of our bodies and minds, a *super*natural naturalism. Whatever it may be in itself, its "intrinsic value" is "manmade" for human ends, and its "beauty" a function of our sensory apparatus. Protecting "nature" is in our own best interests, unless we are ecocentric enough to bequeath "our" world to the next generation of mutants, perhaps irradiated survivors of Chernobyl (who might thrive on pollution and nuclear wastes) or animals bizarrely transformed by gene therapy run amok.

As I reached the eastern side of the Sierras, everything changed. The lush green mountainsides had now become the dry and sere landscape of the high desert. Off in the distance, like a New Atlantis, glowed the high-rise casinos of Reno, suffused with an orange halo from the buildings' nighttime illumination systems. Another feat of techno-magic.

I spent a few days in Reno and Sparks with my friends, then moved on to Salt Lake City, whose Wasatch mountains provide what is perhaps the most awesome stretch on all of I-80, though again the placement of the city and the cut of the road contribute a very humanized aspect to the spectacle, like a Turner painting. If the ill-fated Donner party could be said to have suffered from a *lack* of technology, doomed as they were by slow locomotion, lack of electronic communications, inadequate maps

and weather information, low-tech food storage and preservation, absence of railroads and cities, and insufficient ways of keeping warm, then
my own near-catastrophic hour in a terrifying whiteout episode at seven
thousand feet in the Rockies outside Laramie was the result of advanced
technology that was not quite advanced enough. When violent winds
began to blow the snow from the mountaintops almost horizontally across
the highway, the marvelous roads became virtually invisible and treach-
erously icy; my high-tech car was comfortable and warm in the twenty-
degree April air, but direly in need of X-ray vision or radar. On the side
of the road was a disabled juggernaut, a cross-country semitrailer, now
overturned from excessively daredevil driving. The other cars I could barely
see despite their generally ample lights. All the normally enabling technologies made it possible for me to zoom into incredible danger with the
insouciance and intrepidity that technology so often breeds. (The failures
of modern technology have been cruelly charted by David Ehrenfeld in
The Arrogance of Humanism.) With parched mouth and pounding heart,
I was nearly rear-ended by another semitrailer that barrelled out of the
curtain of snow into my rearview mirror as I crawled along at 25 mph,
managing to swerve aside at the last possible second.

Then suddenly, as though turned off by a switch, the whiteout lifted,
and I found myself in crisp bright sunlight. The next day, approaching
Lincoln, Nebraska, with winds again howling but sun shining brightly, I
was confronted by another sort of techno-misadventure, this time a blind-
ing dust storm that probably covered hundreds of square miles, the result
of recently plowed fields and drought-like conditions—fortunately not very
serious compared to the adventure of the day before. For me the progno-
sis was plain: not a return to hunting and gathering—a never-never land
of innocence and stasis—but a more and more refined technology. Technology is a metaphor for evolving human life, with consciousness as its
blueprint. It is no more reversible than consciousness itself.

After a night outside Omaha, I was returned to bliss once more: as
my estimable radio's scanner sampled every AM/FM broadcast receivable
along I-80 in Iowa, I soon discovered the excellent station operated by
Iowa State University, audible across most of the southern part of the
state. I decided to give the CD player, which had provided exemplary service, a much-earned rest. It was late Saturday morning, and the Metropolitan Opera was about to broadcast *Die Walküre,* conducted by James

Levine, my favorite part of *The Ring*. With Bruno Walter, Lotte Lehmann, and Lauritz Melchior necessarily delivering a shadow performance deep inside my head and heart, I was moved nonetheless by the power of Levine's introductory thunderstorm, and I was knocked quite flat by the heldentenor virtuosity of Plácido Domingo as Siegmund. As I reached the Quad Cities area and began the crossing of the Mighty Mississippi—with due regard for the river-defining technologies of Mark Twain, T. S. Eliot, and the Army Corps of Engineers—signs of spring were definitely in evidence, buds were opening, the air was warming, and Sieglinde was singing the most rapturous passage in all of *The Ring*: "Du bist der Lenz / nach dem ich verlangte / in frostigen Winters Frist."

As tears of ecstasy again began to flow from the easily unraveled strands of my subtile knot, like Molly Bloom saying yes, I experienced a powerful moment of assent to a newfound identity: fighting postmodernism all the way, I had nevertheless to acknowledge that I was indeed, after all . . . *a cyborg*! Take away technology, I realized, and I would cease to exist. From my daily traveler's lunch of frozen yogurt spurted forth from machines at McDonald's to my Visa card swiped through roadside gas pumps, I began to review the adventures of the previous three weeks. And while I can't provide a summation as resoundingly scriptural as Eliot's "What the Thunder Said," I think I can venture a little homily called "What the Car Stereo Said."

To a greater or lesser degree, I therefore affirm, everything human is technological. Everything human is anthropocentric as well. "Ecocentric appreciation of nature" may have a disinterested honorific air (like "for the greater glory of God"), but if the "nature" that we "appreciate," like Wordsworth's Alps, is largely produced by our psychobiological constitutions (Wordsworth called it "Imagination"), then appreciation of nature (and everything else) is an essentially anthropocentric subjectivity. Because, if *we* ceased to exist, the "majestic Sierras" would cease along with us. *Something*, presumably, would remain (e.g., the universe dealt with by the physical sciences), but it would be neither peaks nor blobs, neither majestic nor "serrated," all requiring a sensibility and a "point of view." From the point of view of the universe (which has no point of view), to name it is to misname it, because the act of naming makes it what it is *only for us*. Without nature, no humans; but without humans, no "nature." If this is true, then what can it mean to be "ecocentric"? The motto of the journal *Interdisciplinary Studies in Literature and Environment (ISLE)* reads, "When humans study nature, it is nature studying itself." But it might just as well read, "When humans study nature, it is humans studying themselves."

As I pressed into Illinois, now only a few hours from home, act one of *Die Walküre* was drawing to its close: Plácido Domingo, in the guise of Siegmund, shouted his "Siegmund heiss' ich / und Siegmund bin ich!"— a feat of the most glorious vocal technology. He pulled his sword out of the tree with that unbearably potent cry of "Nothung! Nothung!" leaving multitudes of technologically equipped people all over the world gasping at their stereos in Wagnerian delirium. Sieglinde, as my libretto so finely puts it, "throws herself passionately on his breast" while "he draws her to him with passionate fervor." To worldwide gooseflesh, the orchestra played its wild and frenetic coda while the lovers embraced. As the curtain fell at the Metropolitan and the audience went berserk almost a thousand miles to the east, I thought of the way in which everything suddenly falls into place at the end of *To the Lighthouse*—and I felt that I too had had my vision.

CHAPTER ELEVEN

Full Stomach Wilderness and the Suburban Esthetic

> These wild things, I admit, had little human value until mechaniza-
> tion assured us of a good breakfast . . . When we see land as a com-
> munity to which we belong, we may begin to use it with love and
> respect. There is no other way for land to survive the impact of
> mechanized man, nor for us to reap from it the esthetic harvest it
> is capable, under science, of contributing to culture.
>
> ALDO LEOPOLD, FOREWORD TO
> *A Sand County Almanac*

I WAS INDUCTED INTO THE environmental movement in the early seventies as a result of an idiotic move to a seemingly idyllic farm located
only fifteen miles south of the steel mills of Gary, Indiana. In those days I
was not alone in being innocent of the fact that pollution traveled not just
fifteen miles but fifteen hundred miles and more. But the resulting nightmare, illnesses both bodily and psychological, transformed my life and re-
cruited me into the ranks of the ecologically committed. I wrote about
and waged campaigns against air pollution and the general depredations
of corporate environmental destruction. After an escape to the northwest
suburbs of Chicago in North Barrington, Illinois, I continued my ac-
tivism, this time not only with regard to industrial pollution but to pesticide spraying for mosquitoes, leaf burning, water contamination by runoff into wells and aquifers, and so forth. The village trustees hated my
guts. But despite this ecological commitment, I never identified with ter-
rorist types—animal rights fanatics who destroyed laboratories and opened
cages on family farms, ruining multiple lives in the process, or Earth First!
types who spiked trees climbed by actual human beings who are maimed
for life. I found Dave Foreman's remark that he would sooner shoot a

man than a grizzly rather far over the top (though Foreman, like other radicals from the sixties and seventies, has since morphed into a pussycat). To me, these were self-involved narcissists, no better than the bombers of abortion clinics and murderers of physicians. With friends like these, who needed enemies?

But it's probably safe to say that even extremist types have done some good in jump-starting the reforms of society. The trouble for me, however, is that relentless one-string activists approach too closely to religious fanaticism, are too certain of pious absolutes that tomorrow will be seen as personal pathology. Ecological Jerry Falwells are just not my cup of tea. So when it comes to wilderness, I'm suspicious of uncompromising purities—even when they come from Thoreau or Leopold. I've said and written before that I don't believe anything human can ever be other than anthropocentric—and that biocentrism is just anthropocentrism in pious drag, like Jerry Falwell telling us what God wants, a God who always turns out to have the atavistic brain of Jerry Falwell. Of course, there are different varieties of anthropocentrism—and some are more benign than others.

So the sentimental extolling of wilderness found in a book like Max Oelschlaeger's *The Idea of Wilderness* never really appealed to me, and his golden age view of hunter-gatherers seemed preposterous. Still, wilderness was a fairly abstract thing during my thirty years in Chicagoland's prairies and it was not until my move to Tucson in 1998 that it became a concrete reality intersecting my daily life. If I was skeptical then about wilderness equals pure, society equals impure, I am now a total nonbeliever. I live on a ridge in the foothills of the Santa Catalina Mountains north of Tucson where every day and every sunset are spectacular. I can testify to the fact that Tucson is surrounded by vast areas of wilderness, most of it quite inaccessible, despite the foot trails that afford entry into the mountainous areas closer to town. Between my home northwest of the city and the Pinal County seat of Florence, southeast of Phoenix, lies a forty-five-mile stretch of backroads desert that is unsettled enough for me finally to have bought a cell phone, so that in the event of an auto breakdown my bleached bones would not be discovered next day by a scouting party attempting a rescue.

The Coronado National Forest spreads its discontiguous immensity all over southeast Arizona, and there are vast national parks and Indian reservations as well that occupy much of the state. Between Phoenix and Flagstaff and Prescott lie additional uninhabited and majestic vastnesses. The eastern United States may be packed, dense, and built out with habitation, but once you cross the Mississippi, the wide-open spaces are not purely mythic. Much if not most of this area will never be hospitable to settlements, and the environmental mentality that bit by bit is spreading its influence offers greater and greater resistance to wildcat development, even if it's too late to save Phoenix. Tucson has been remarkably resistant to freeways and massive urbanization, so even today it has paradisal qualities, plunked as it still is in the middle of wilderness, though a realist view would prepare for its inevitable San-Diegoization as the pace of building speeds up along the beckoning corridor to Phoenix. But what's left is immense nonetheless.

It has become a truism that the wilderness is a modern invention, an esthetic object, a rhetorical device, that didn't exist for people who inhabited it when there was nothing else. When I drive around the Tucson suburbs and view the astonishing beauty of the mountains or hike in the trails of the Catalinas, I am struck by the fact that it is often suburban development that has freed up, even created, the breathtaking vistas for which Tucson is famous. The fabulous sunsets seen from Gates Pass Road and the shimmering beauty of monsoonal fogs on the Catalinas have been made visible for us by the accomplishments of pampered, technocratic, impure bourgeois like ourselves. In contrast, to walk through the ruins of the Hohokam Indians (A.D. 500–1500) in Catalina State Park is to marvel at a life that appears almost impossibly brutal amidst the dry, burning summer heat and seeming absence of shade, water, food, and websites. Could there have been substantial periods of "quality time" during which these environmentally challenged people sat around and exclaimed over the scenery like us, while shamed by their good luck at being mostly skilled farmers instead of full-time hunter-gatherers? Or, too oppressed for leisure, did they instead secretly harbor inchoate longings to become twenty-first century bourgeois—connecting with enough "nature" at the Arizona–Sonora Desert Museum followed by high tea at the Tohono Chul Tea Room—so that they could really begin to enjoy the wonders (instead of the obstacles) of where they lived?

I contrast the fears of that primitive habitat with the undeniable appreciation of their surroundings displayed by the residents of my community of SaddleBrooke, nestled in the Catalina foothills about twenty-five miles from downtown Tucson in the outermost fringe of suburban development. Retired into the leisured life of affluent bourgeois, they seem to have world enough, food enough, and time, given our extended lifespans, to appreciate the value of the "unspoiled," of the "domesticated sublime," as William Cronon aptly describes it in his definitive essay "The Trouble with Wilderness."[1] I marvel at the paradox: as suburban development spreads into the wilderness, it both destroys and produces it at the same time; as the wilderness recedes farther and farther, it becomes an object of contemplation to be valued rather than feared. Only then do mountain lions and bobcats acquire an autochthonous beauty that fills us with Prufrockian guilt about disturbing "their" universe. How beautiful could it all possibly have been to the Donner party trekking from the Midwest to the West Coast without sports energy bars, SUVs, cell phones, or coats of Polartec fleece from Lands' End? Those who survived needed to be us to appreciate the beauty that sheer surviving made impossible for them to see.

Nowadays, some of the most remote and undeveloped wilderness can be explored and enjoyed through the prostheses afforded by contemporary technology, from motorized vehicles (a blessing as well as a curse), water-purification equipment, and space-age clothing to freeze-dried food, propane cookers or sun ovens, and cell phones, conjoined with the reassuring sense that one has a home elsewhere, a safe haven, when supplies run out. When all else fails, helicopter rescues are yet another twenty-first-century entitlement of Everyman. Thus, the social and the technological give us more and more of a highly valued wilderness they seem to be pushing farther and farther away. (I refer the reader to Cronon's essay for an exemplary account of the myth of purity and pristinity that undergirds the putative sanctity of the untransgressed.)

An open-ended negotiation between development technology's creative and destructive forces may be the only resolution of what qualifies as cohabitation with "the natural." And even what constitutes creation and destruction is hardly self-evident or clear. As I emerge from the supermarket to sensational mountain vistas or sip beer with my pals at the outdoor tables of my favorite brewpub against the panorama of Pusch Ridge, I feel that these wonders have been produced for me by the very forces I have given others money to suppress. What counts as creation or destruction, however, is based on values springing from ever-changing human subjectivities, with their subterranean desires and unexpressed ideologies. (Yesterday's sickies are today's culture heroes—and vice versa.) The only plausible moment of pristine innocence must have been the microsecond before the Big Bang—and even that looks to be a bit suspect if it was capable of producing a corrupted something out of a pristine nothing.

Despite the environmental setback of a president who comes off as a reincarnated Saudi oil prince, underwriting corporate greed as the well-being of American life is compromised by handouts to industry, we know from the Reagan era that a period of ecological reaction is destined to set in—and of course it is happening even now. But first, more damage than necessary will evidently be done, leaving more to be corrected afterward. Gas mileage, air-conditioning SEERs, alternate energy subsidies, etc. have all had recent setbacks from Republican rapacity and shortsightedness (not that the pusillanimous Democrats are so much better). But human life requires limitations to survive urban and suburban development—and once we are personally developed enough to become bourgeois, aesthetic needs kick in as well, which increase the desire for limitations. Valuing nature is a middle-class enterprise, and even those putative despisers of American middle-class life who claimed to be rescuing the world for the rest of us—like Bernardine Dohrn, Sara Jane Olson, Ted Kaczynski—turned out, in one way or another, to be products of bourgeois amenities. It's only after a full stomach has been assured that we are suddenly open to a whole spectrum of salvific epiphanies, not the least of which is the fantastic realization that the spotted owl, *c'est moi.*

CHAPTER TWELVE

Coetzee's Postmodern Animals

"THERE USED TO BE A TIME when we knew. We used to believe that when the text said, 'On the table stood a glass of water,' there was indeed a glass of water, and a table, and we had only to look in the word-mirror of the text to see them both." So remarks Elizabeth Costello, a novelist invented by J. M. Coetzee for a talk on realism at Bennington College in 1996 and destined to reappear in Coetzee's novella, *The Lives of Animals*. In the earlier incarnation she has flown from Australia to Appleton College in Massachusetts to receive a literary award that prompts her to discuss the matter of realism in literature. "But all that has ended," she continues, "The word-mirror is broken, irreparably, it seems . . . The words on the page will no longer stand up and be counted, each saying, 'I mean what I mean!'" Disintegration has set in even further: "The dictionary, that used to stand beside the Bible and the works of Shakespeare above the fireplace, in the place occupied by the household gods in pious Roman homes, has become just one code-book among many." The ramifications for a writer are dire: "There used to be a time, we believe, when we could say who we were. Now we are just performers, speaking our parts."[1]

Risky as it may be to use fictional words as if they expressed the sentiments of their creator, we're not going to get very far in connecting with this, or any, author without trying to establish a point of view. And with someone as slippery as Coetzee (a performer speaking his part?), we need all the help we can get. Coetzee sets his new lives of animals into the existing framework of his Bennington talk about realism, retaining not only Elizabeth Costello but her son, Appleton College, and the idea of a public lecture, but the substance has been drastically changed. Costello's diffidence about the correspondence between words and things—a legacy of Coetzee's years in the United States as a linguistics graduate student during the period of structuralism and deconstruction—mitigates somewhat her powerful moral assertiveness in the earlier version but functions more quietly in the later where, instead of literature, she lectures on animal rights.

This diffident assertiveness is perhaps Coetzee's most salient characteristic as postmodern author for whom playful realism has necessarily replaced oracular prophecy. Yet "postmodern" can be misleading unless a distinction is made between the two postmodernisms: of literary technique and human personae. While on the one hand Coetzee's *prose* is increasingly lean and straightforward, not "experimental" or tricksy (although his early *In the Heart of the Country* comes off as a sometimes tedious cross between Woolf's *The Waves* and a Faulkner novel), on the other hand his literary *persona* reflects the philosophical modesty that comes from living in a world with radically damaged foundations of belief. Realism is a way out, a means by which the postmodern literary persona can seem to avoid postmodern techniques: "Realism has never been comfortable with ideas," he remarks in his own voice in the Bennington lecture, sharing Costello's fictional point of view. "Realism is premised on the idea that ideas have no separate existence, can exist only in things. So when it needs to debate ideas, as here [in the Elizabeth Costello story invented for the lecture], it is driven to invent situations—walks in the countryside, conversations—in which characters give voice to contending ideas and thereby in a certain sense embody them. The notion of embodying is cardinal" (65). It's worth adding that the fiction Coetzee has invented to present his presumed endorsement of realism is filled with echoes of other writers at the same time that we hear a complaint about rehashing the classics; it expresses the view that language, timebound and conventional, writes us (rendering genuinely creativity almost impossible) even as one of the characters expresses a preference for making it new; and it ends with Costello's son contesting his mother's esthetic and extolling the creative miracle of the Romantic imagination while rejecting realism's "smelly underwear."

But Coetzee's postmodern persona has been throwing curveballs for a long time. In an oft-quoted 1984 essay about his time as a graduate student in English at the University of Texas in Austin (where he arrived from his native South Africa in 1965), he wrote: "If a latter-day ark were ever commissioned to take the best that mankind has to offer and make a fresh start on the farther planets, if it ever came to that, might we not leave Shakespeare's plays and Beethoven's quartets behind to make room for the last speaker of Dyirbal, even though the last speaker might be a fat old woman who scratched herself and smelled bad?"[2] But don't expect Coetzee to rush out to make it with the fat lady. For with him, what the left hand giveth, the right hand taketh away.

Coetzee started out in mathematics, moved on to linguistics, and found his niche in fiction. In an interview in *Salmagundi*, he remarks, "Mathematics is a kind of play, intellectual play. I've never been much interested in its applications, in the ways in which mathematics can be set to work. Play is, to me, one of the defining characteristics of human beings." And when asked about the implications of fictional narrative, he replies, "I can't make an exhaustive list, but they do include abandoning the support that comes with a certain institutional voice, the voice of the historian or sociologist or whatever. It entails no longer being an expert, no longer being master of your discourse."[3]

In still other interviews, these notions of play and the inauthenticity, so to speak, of the author keep resurfacing. He speaks of "an awareness, as you put pen to paper, that you are setting in train a certain play of signifiers with their own ghostly history of past interplay," and he wonders whether earlier writers like Defoe felt similarly trapped by history and culture. The contemporary writer is "like children shut in the playroom, the room of textual play, looking wistfully through the bars at the enticing world of the grownups, one that we have been instructed to think of as the mere phantasmal world of *Realism* but that we stubbornly can't help thinking of as the *real*." Though he doesn't want to "sound silly" by talking about the muse, writing is an activity cut off from daily life and propelled by a dynamic of its own, a dynamic that makes the everyday person who happens to be the author not much more of an authority about his writings than the reader who reads them.[4] Unsurprisingly, Coetzee has been criticized for his relative disengagement from politics—unlike his countrywoman Nadine Gordimer—and for not dealing head-on with the turmoil of South Africa, focusing instead on individual consciousness in relation to which South Africa is more background than foreground.

As an academic *malgré lui*, Coetzee comes off at first glance as a model of political correctness. He seems to hold the positions that today's professors are supposed to hold. He is particularly sympathetic to women. In *In the Heart of the Country*, he even writes as a woman. He can be hard on male insensitivity and predatoriness. Yet when pressed for his "correct" opinions by interviewers, he has a way of slipping out of their lassoes, of refining his outlooks and refusing to be tied down. In place of easy-to-define political positions and polemics, which he dislikes, he has come to value (I'll put it in quotes, since I'm creating it as a term here) the "sense of being," a distinctly Heideggerian rejection of time-and-culture-bound social incarnations in preference for living in the body, experiencing one's sentience of the moment without cosmic claims.

This is dramatically revealed in his 1983 Booker Prize novel, *Life and Times of Michael K*. Its protagonist, whom I would characterize as an idiot-savant, wanders around South Africa during an imagined time of what amounts to civil war between the minority whites and the majority blacks, largely oblivious of or uncaring about what is going on except insofar as it impinges on his individual consciousness as a sentient being. Mainly concerned simply to survive in personal freedom, literally and figuratively cultivating his garden without money, home, or health, he sides neither with whites nor blacks (himself presumably black—a fact Coetzee has concealed in the interests of an apolitical focus). Michael K is an illuminating embodiment of Coetzee's muse-driven, existential literariness, his Great Refusal, a hybrid of Melville's Bartleby, Kafka's K and hunger artist, and Camus' stranger (with a bit of Robinson Crusoe thrown in for good measure). In a culminating scene, after Michael K refuses help from the medics in a refugee camp where he has been herded ("What was the food in the wilderness that made all other food tasteless to you?" he is asked) and after he escapes to a wretched, impoverished, freedom, we discover, through a medical officer's internal monologue, how deeply K's mode of being in the world has affected him: "I slowly began to understand the truth: that you were crying secretly . . . for a different kind of food, food that no camp could supply . . . Slowly, as your persistent *No,* day after day, gathered weight, I began to feel that you were more than just another patient . . . Did you not notice how, whenever I tried to pin you down, you slipped away? . . . The garden for which you are presently heading is nowhere and everywhere except in the camps . . . It is off every map, no road leads to it that is merely a road, and only you know the way."[5]

This rejection of social constructions and abstractions in favor of immersion in a uniquely personal sense of concrete being during a time of unbelief (in this instance, the collapse of traditional South African society at the end of Apartheid) is an inveterate feature of Coetzee's own psyche, a feature that accords well with the uprootedness of postmodernism as it animates his fiction and interviews. A reluctantly cerebral academic, he would rather just *be*. Instead, the world of political correctness, of which he is only a half-hearted citizen, keeps flogging him as it did Michael K to reveal which side he is really on.

The Lives of Animals brings to bear on many of the contradictions described above. Presented as the Tanner Lectures at Princeton in 1997–98 shortly before the brouhaha resulting from Princeton's appointment of Peter Singer, author of *Animal Liberation,* to its philosophy department, the presentation is a veritable postmodern hall of mirrors: a fiction writer, J. M. Coetzee, invited to speak philosophically on an ethical problem, instead gives two lectures that are in fact short stories about a fiction writer, Elizabeth Costello, who is expected to talk about literature but who instead gives two lectures on the philosophic subject of animal rights. The characters, locale, and venue are given slightly askew versions of the names of actually existing professors and places, while the narrative that surrounds the fictional lectures is highly critical of both Costello herself and the things she has to say. As if this weren't dizzying enough, the book version is laced with authentic scholarly footnotes by Coetzee documenting his fictional statements about animal lives and rights, and it concludes with brief commentaries by four preeminent thinkers in literature, philosophy, religion, and anthropology, all packed into 122 supercharged pages.

This time around, Costello again visits her son and daughter-in-law at Appleton College, but they are none too happy to see her, knowing that she is a difficult and uncompromising person who will soon attempt to make everyone feel guilty about killing animals and eating their meat. The son, a professor of physics and astronomy, is wary of her tendencies toward emotional blackmail, and the daughter-in-law, a professor of philosophy, finds Costello's views jejune. The story is told with uncharacteristic wit and irony, rhetorical modes that Coetzee has customarily avoided but that here serve as intentional feints to throw the reader off the trail that might identify the author's point of view (which he himself may not yet have identified). In her lecture at the college, Costello informs her audience that she will skip a concrete recitation of the horrors animals suffer through human abuse and will confine herself to more general issues. Comparing the farming of animals for meat and lab experiments to the slaughter of the Holocaust (with analogies to the gold fillings and skins for lampshades culled from the corpses), she remarks: "Let me say it openly: we are surrounded by an enterprise of degradation, cruelty, and killing which rivals anything that the Third Reich was capable of, indeed dwarfs it, in that ours is an enterprise without end, self-regenerating, bringing rabbits, rats, poultry, livestock ceaselessly into the world for the purpose of killing them."[6] She attacks Plato, Aquinas, Descartes, and Kant for their views of animals as machines designed for human gratification and makes an all-out assault on reason (which is used to justify animal abuse) as a socially self-interested human construct (25): "Of course reason will validate reason as the first principle of the universe—what else should it do? Dethrone itself?" (But her daughter-in-law will respond: "Human beings invent mathematics, they build telescopes, they do calculations, they construct machines, they press a button, and, bang, *Sojourner* lands on Mars, exactly as predicted . . . Reason provides us with real knowledge of the real world" [48].)

Costello, like Aldo Leopold, Roderick Nash, and other ecological thinkers, wants to extend the range of personhood (in varying degrees) to include animals and is militantly opposed to a Cartesian cogito that favors rational consciousness over animal sentience. "To thinking, cogitation, I oppose fullness, embodiedness, the sensation of being—not a consciousness of yourself as a kind of ghostly reasoning machine thinking thoughts, but on the contrary the sensation—a heavily affective sensation—of being a body with limbs that have extension in space, of being alive to the world" (cf. Michael K). You can't feel full of being when you are confined to a pen with other animals by keepers who refuse to "think themselves into the place of their victims" (33). Costello's claim for human ability to think oneself into the Other is based on this intuition of a shared sentience of being, but its assumptions are fallacious and weak, resembling Aldo Leopold's "thinking like a mountain," which turns out to be thinking like a person thinking about a mountain. Believing that she *really can* share animal feelings, she quotes with disapproval the philosopher Thomas Nagel on the futility of trying to think like a bat: " 'Insofar as I can imagine this [how it feels to have webbed feet] . . . it tells me only what it would be like for *me* to behave as a bat behaves. But that is not the question. I want to know what it is like for a *bat* to be a bat'" (31). Like most of Costello's claims, there is a certain emotional power in what she has to say on this subject, but it is the power of human desire rather than the power of truth, a distinction she seems unable to make. There is a danger of narcissistic ruthlessness in this form of presumed human sympathy, and when it is carried to its extreme, when one "speaks for" God or the state or the race—or animals—(i.e., when one's puny human consciousness puts on world-historical or cosmic robes à la Pat Robertson and Jerry Falwell), there may very well be a holocaust lurking in the wings (one thinks of Conrad's Mistah Kurtz). Her son is fearful she will be asked why she doesn't eat meat and that she will give her customary antinomian answer (which chills his blood): "I, for my part, am astonished that you can put in your mouth the corpse of a dead animal, astonished that you do not find it nasty to chew hacked flesh and swallow the juices of death-wounds" (38).

When her son asks whether she would put a jaguar on a soybean diet, she says no, "Because he would die. Human beings don't die on a vegetarian diet." But he replies, "No, they don't. But they don't *want* a vegetarian diet. They *like* eating meat. There is something atavistically satisfying about it. That's the brutal truth" (58). Other fictional respondents to Costello make equally harmful points, ranging from still more defense of reason: "There is no position outside of reason where you can stand and lecture about reason and pass judgment on reason" (48) to the distinction between a human and an animal death. Since animals, unlike people, do not fear death, "For that reason . . . dying is, for an animal, just something that happens, something against which there may be a revolt of the organism but not a revolt of the soul . . . It is only among certain very imaginative human beings that one encounters a horror of dying so acute that they then project it onto other beings, including animals" (63).

Not "a revolt of the soul"! The distinction is startling in its brilliance— and when Costello is pushed against a wall, unable finally to offer convincing reasons for her aversion to animal killings, she replies that ultimately she wants to save her soul. But what is this soul, in any case, but human subjectivity, projecting its desires upon the canvas of a silent universe amenable to infinite interpretations? Costello's performance is eloquent and powerful but in the final analysis unpersuasive.

The replies to Coetzee's two lectures are all of great interest, but it is

Peter Singer's that most directly addresses the issues in question. Singer's landmark book, *Animal Liberation* (first published in 1975), was principally concerned with the pain and suffering we inflict on animals, particularly in the needless mass production of meat by means of farm factories. His accounts of the torturous process whereby veal calves are penned, constricted of movement, and deprived of proper nutrition to make them anemic was horrifying, and his description of chicken raising no less alarming. He recommended vegetarian diets, railed against "speciesism" (which ignores the complex lives of other species), and detailed the extraordinary waste of natural resources involved in feeding calves twenty-one pounds of plant protein to produce one pound of animal protein for humans.[7] (Even now, the American meat industry is moving into China, cutting down forests for grazing, and introducing Western diets to a people largely free of the diseases connected with the highly saturated fats of animal-based foods.) There are so many devastating byproducts of a carnivorous diet (destruction of trees needed for oxygen and erosion control, bad human health, pollution from mountains of excrement, production of methane gas, denuding of prairies from grazing, etc.) that a good case can be made for limiting it. Yet even Singer is somewhat turned off by Elizabeth Costello's antinomian retreat to "feeling" as arbiter of moral action. In a little short story of his own that he wrote as comment upon *The Lives of Animals,* Singer remarks through his fictional stand-in, "When people say we should only *feel*—and at times Costello comes close to that in her lecture—I'm reminded of Göring, who said, 'I think with my blood.' See where it led him. We can't take our feelings as moral data, immune from rational criticism" (88–89).

Elizabeth Costello's selective derogations of reason and retreat to soul-saving feeling ignore the fact that not only have standards of behavior toward animals (apart from industrial farming) improved during the past century (try harming suburban raccoons, deer, or even Canada geese and see what trouble you're apt to get yourself into) but that—to repeat words I have used on prior occasions—to be alive is to be a murderer. Anyone who has witnessed five lions tearing apart a living zebra on the Discovery Channel or PBS can hardly be sentimental about the benignity of animals toward other animals, nor is there any way to stay alive without killing something else for sustenance. Doubtless, the more complex the consciousness, the more criminal the cruelty toward it—and there is plenty of room for improvement. But the real original sin would seem to be life itself, which continues only through the exercise of slaughter and destruction. What are animal rights but the obverse side of the coin of animal sacrifice, an attempt to expiate the guilt of simply being alive by making a show of relinquishing a little power? Whether this is a "good thing" is another matter. A totally scrupulous moral consciousness might want to consider suicide, the ultimate gift to Others—because "alive" and "innocent" can't be put into the same sentence except as oxymoron.

In Coetzee's latest novel, *Disgrace*,[8] David Lurie, a libidinous white South African professor, is unapologetic after he seduces an all-too-willing young student, who later charges him with sexual assault. Forced to resign unless he recants and contemptuous of the political correctness that lies behind the treatment he receives from his university, he departs from Cape Town and goes to live with his convention-flouting lesbian daughter, who owns a small farm in a post-Apartheid countryside now increasingly dominated by majority blacks. After initial feelings of superiority to much of his new surroundings, he begins to work in an animal shelter run by his daughter's neighbor and gradually finds himself humbled by pathos and tender feelings toward the dogs being put to death for lack of any takers. In a shocking episode, the daughter is raped by three black young men, acquaintances of her black, formerly socially submerged, neighbors; her own dogs are shot and her father is set on fire but survives in mostly good shape. Lurie, who loves his daughter, is fearful of AIDS and pregnancy, but the daughter, who indeed becomes pregnant, accepts it as an inevitability of the new order and decides to keep the child. The novel ends with Lurie tenderly escorting doomed dogs to their death.

What is one to make of all this? There is, after all, no unqualified spokesperson in either set of fictions. Costello, who emerges as the impassioned protagonist, is so widely and persuasively criticized, her arguments so shaky, that it is hard not to wonder what Coetzee is up to in placing her center stage. And who is the spokesperson for *Disgrace*? Is it Lurie, whose political incorrectness is treated as abrasive (yet who is nonetheless rather winning in the current atmosphere of moral blackmail), who functions for most of the novel under a cloud of disapproval, but whose "conversion" is admittedly regarded with sympathy? (Conversion, however, to little more than a compassionate dogzapper.) It's surely not his daughter, whose righteousness is off-putting and whose willingness to be raped and otherwise violated for the new social order is less than charming. Is the younger generation necessarily wiser than the older? Are we really to think that black thugs are more virtuous than white ones?

What, finally, can a postmodern author, an academic intellectual, an Afrikaner with a mission more cosmically esthetic than locally moral (and because of it, criticized by his countrymen) seem to be telling us? Coetzee's flirtations with political correctness have a nervous and self-canceling quality. He recognizes the validity of many of its claims but he is unable to avoid supplying damaging counterarguments through his wide spectrum of characters, who collectively see through everything. Are vegetarianism or gentle canine euthanasias going to snatch us from our original sin of being alive? Will accommodating ourselves to the inevitabilities of "progressive" history save our always-already damnable souls? Does immersing ourselves in a bodily sentience of being rescue us from trendy moral ideas that have a nasty way of looking corrosive when their currency expires? What's a postmodern person—author, reader—to do?

Coetzee seems to have taken on the only plausible role for a fin de siècle skeptical wise man who sees too much for his own good. As a sensitizer rather than a doctrinizer, he has limned with skill a multiplicity of human possibilities, veering toward a handful of what seem like desirable choices but always acutely conscious of the treacherousness of history and the self-deceptions of human consciousness. His point of view, his moral stance, are hard to determine, although one senses a tentative drift maintained with sails never fully rigged. Looking out for animals, giving new social orders their due, entertaining the claims of Others, remaining open but unbedazzled—this is consciousness-raising without doctrine, or perhaps a fine-tuning of the stethoscope by means of which a postmodern author listens in—with some anxiety—on the murmuring heart valves of humanity.

PART TWO / "Nature" and Evolution

CHAPTER THIRTEEN

My Science Wars

Although it was in the early eighties when I began to feel a growing disaffection with the radicalized academic left, a decisive nausea-inducing body blow was administered by the *PMLA* of January 1989. In that infamous issue appeared a letter signed by twenty-four feminist academics attacking the eminent Shakespeare scholar, Richard Levin, for "Feminist Thematics and Shakespearean Tragedy," which had appeared in *PMLA* the year before. Levin's essay, the work of a well-tempered, open-minded, and liberal supporter of many radical reforms in the academy, was a penetrating critique of the feminist identity politics that were trying to wrest Shakespeare into the persona of a feminist scourge of sexism and patriarchy. Levin's temerity in taking on the distortions of identity politics was threatening enough, but his Enlightenment style and Swiftian wit, which he brilliantly deployed to dismantle the tendentious and self-serving hermeneutics of the "dictatorship of virtue" (a phrase I borrow here from Richard Bernstein) that constitutes avant-garde millenarianism in the U.S. academy, served to further inflame the hypersensitive skins of his holier-than-thou critics, who brook no criticism because they are already in possession of the absolute truth they deride in all other cases. Three gems from that rather lengthy letter beg to be quoted, although they cannot begin to convey the noxiousness of the whole:

> We argue that gender difference is a historically specific cultural construct with diverse forms and representations and damaging consequences for characters in plays, subjects in the Renaissance, and for us—and Levin—today . . . The view that "science" and "rationality" can comprehend "complex factors in human development" without the messy intrusion of "gender and ideology" is an Enlightenment dream, long since turned to nightmare . . . We wish to know why, in view of the energetic, cogent, sophisticated theoretical debate that is

currently taking place within and among schools of Renaissance criticism, *PMLA* has chosen to print a tired, muddled, unsophisticated essay that is blind at once to the assumptions of feminist criticism of Shakespeare and to its own. (77–78)

If ever a critical essay had been radically (in all senses) misrepresented in the interest of identity politics, Levin's piece on feminist thematics in Shakespeare was certainly a locus classicus. Far from being tired and unsophisticated, Levin wrote then and continues to write now with vitality, wit, insight, and precision. This misbegotten attempt to bad-mouth and silence him was a performance of the radical academy at its "final solution" worst. (Academic final solutions don't require gas chambers, just lots of gas.) And when I saw the moral character of the opposing sides, I asked myself then as I regularly do now: if forced to make a choice in delegating power, into whose hands would I be willing to put myself? In cases like this, the question is hardly even worth asking.

Only a few years later I was invited to contribute an essay to a collection dealing with Gerald Graff's pedagogy of "teaching the conflicts." Rereading a number of his essays that formed the basis for the collection, I was struck by Graff's democratic and egalitarian insistence that postmodern literary theory, far from being "obscure, technical and abstruse, and therefore too advanced or esoteric for the average college or high school student of literature," could be made perfectly intelligible and accessible because "all teaching involves popularization." Yet when well-educated critics like David Lehman and Robert Alter ventured to write books highly critical of the avant-garde canon, Graff brushed them off as "pretty bizarre" and "totally ignorant." And the inevitable question for me was, "Easy for students but impossible for David Lehman and Robert Alter?" "One begins to fear," I added, "that what Graff really wants is not conversation or diversity, but *absolute conformance and identity.*"[1]

Now, in 1996, I witness the fortress mentality of the culture wars reaching some sort of apogee in its latest phase: *Science Wars.* This, the title of the spring/summer issue of *Social Text,*[2] was generated as a response to *Higher Superstition: The Academic Left and Its Quarrels with Science,* by Paul R. Gross and Norman Levitt, published in 1994 and still producing a lot of heat.[3] Gross, a biologist at the University of Virginia, and Levitt, a mathematician at Rutgers, are the designated inheritors of the mantle of Richard Levin, and the sixteen contributors to this volume are more or less the brothers and sisters of the twenty-four signatories of the protest letter to *PMLA*, though the effect is not quite so toxic, despite a few pretty dotty moments. Indeed, Gross and Levitt, as scientists who speak the issues and languages of the humanities with exemplary sophistication as men of wide-ranging culture, write an Enlightenment prose rivaling Levin's in its clarity, insight, and barbed wit. And its relatively measured coolness, like Levin's, has so enraged some of their respondents in science studies that *they* have called Gross and Levitt "hysterical." Add to all this the now notorious parody of science studies (unwittingly accepted by the editors) by Alan Sokal, a physicist at New York University, which rounds out this issue of *Social Text*, and you've got the makings of a postmodern reign of terror. "Off with their heads!" resounds from every side, and what should be sage metaphysical lucubration begins to sound like the pandemonium set off by the rape of the lock.

Although a number of reviewers were scandalized that Andrew Ross and the other editors behind *Science Wars* were unable to recognize Sokal's contribution as a hoax—and a few overheated critics on the right seemed to think it portended a takeover by the barbarian hordes and the collapse of Western civilization as we know it—there is no obvious reason why Ross and the others *should* have recognized it, since it sounds remarkably similar to much of what is written in cultural studies. (That was the point, wasn't it?) Nor should one assume that the rest of this issue of *Social Text* is an unmitigated disaster—because it's not. Although it has its share of loony tunes, as a specimen issue of an academic quarterly journal it could probably be defended as better than average. It's certainly not uninteresting. Still, even the best of the essays reveal that a substantial quantity of what passes for sense in cultural studies is indistinguishable from nonsense—and when the nonsense is the *right* nonsense, business can go on as usual, as Sokal's clever parody admittedly suggests. If this means there is something rotten in the state of cultural studies, so what else is new? There is plenty of rottenness elsewhere as well. Unfortunately, the uproar provided more fuel to the fortress mentality, and Ross and Bruce Robbins wrote replies that won't add much to their reputations. Stanley Fish, the arch-jokester/trickster of academe (who just happens to be executive director of Duke University Press, the publisher of *Social Text*), wrote a sophistically defensive (one is tempted to say "bankrupt") letter to the *New York Times,* complaining—of all things— that Sokal's joke wasn't funny. So again, what else is new?

Levitt and Gross's book impressed me with its liberal point of view and its defense of the basic reasonableness of science, even while acknowledging the errors and evils that flow from science's complicity with unbridled capitalist technology. That their critics denounce them as neoconservatives hardly carries much weight, since similar critics denounce Sokal as well, even though he taught math in Nicaragua under the Sandinistas and would appear to have impeccable leftist credentials. Nothing except absolute belief in the latest radical doctrines (which will be supplanted tomorrow by new ones) can placate a Left that simultaneously believes there are no foundations for believing anything. Most of the quarrel between the two sides, however, stems from a misunderstanding that is reflected over and over again throughout the essays in the *Science Wars.* Levitt and Gross are mainly concerned with the intellectual processes that lie behind the procedures of the sciences, that is to say, the type of rationality— whereas their critics are concerned with the undeniably catastrophic consequences that have followed from many scientific discoveries (while playing down the stunning benefits that have made us who we are, e.g., pampered bourgeois with the leisure to take sides in the science wars over latte, instead of spending the day gathering sticks for cooking dinner). The critics see the uses of the sciences as "socially constructed," as well as the choices of projects to be funded and the directions in which such fundings push research. These criticisms seem quite well founded, and Levitt and Gross are generally in agreement with them. But however faulty Western "reason" may be, they nevertheless remind their critics that penicillin works just as well in Third World countries as it does here, so the "reason" that lay behind its discovery must have a genuine connection with the natural world and not be just a socially constructed convention of Western capitalist patriarchy.

Among other things, Levitt and Gross object to the notion that feminist science employs a different "reason" from male science. When Evelyn Fox Keller in her biography of Barbara McClintock accuses biologists of using figures of speech that betray their masculine predispositions, like the term "master molecule" for DNA, Levitt and Gross reply that "DNA as 'master molecule' is shorthand for 'initial information source,' nothing more; it carries no implications of 'dominance.'" Praising McClintock's work, they add that "there is no convincing mark on it of femininity, as McClintock herself was the first to insist. Her closeness to the experimental material, her willingness to 'listen to it,' is characteristic of the work of some scientists and less so of others. There are no data suggesting that women scientists display the characteristic, in general, more often than do men" (141–42). One of the more absurd essays in *Science Wars* goes so far as to examine figures of speech in Levitt and Gross's book itself in order to prove that they are just male sexists and can consequently be ignored as unreliable. But, again, if one were forced to put oneself into the hands of Sarah Franklin, its author, or those of Levitt and Gross, no sane person would choose to be at the mercy of what passes for "reason" in Franklin. There may be lots of reasons, but some are more reasonable than others.

Throughout their book, Levitt and Gross point out that even the most fanatical deep ecologists, even the most outraged AIDS protesters, rely on the reports and discoveries of the sciences about species, ozone, the greenhouse effect, AZT, plate tectonics, the age of the earth, etc., "issues that would be unknown and unknowable but for the accomplishments of professional science" (162). Knowledge of this type is not regarded as socially constructed even by protesters but as the best information available about the "real" world. It's not as though Levitt and Gross believed that the sciences know the noumenal essence of the universe, but they do believe there is a "real" connection between the descriptions offered by the sciences and the way the universe is. That the act of knowing mediates between knower and known and is thereby perspectival is not the same thing as saying that nothing valid can be known about anything. "We simply observe that science is, as all the world's experience clearly tells us, overwhelmingly the best trick we so far know for getting the upper hand against disease. And we *know* that the politicized, overtheorized 'criticism' that is our subject offers nothing at all in that direction. Its main effect has been to assure aspiring cultural critics that they can play a significant role in combating AIDS without having to do anything so tiresome as, for instance, abandoning the joys of lit-crit for careers in medicine or molecular biology" (196).

Andrew Ross does not perform very well in *Science Wars*, neither as editor providing an introduction nor as contributor. His "A Few Good Species," with its scattershot critiques of every passing whim and his congenital one-upmanship finally come to naught, a characteristic failing of excessive energy working on too little substance (or too much insubstance). In his general introduction, he gives his blessing to the blurring of the distinction between theoretical and applied science that permeates this volume as a whole, a distinction that a few contributors go so far as to disallow altogether. "Once it is acknowledged that the West does not have a monopoly on all the good scientific ideas in the world, or that reason, divorced from value, is not everywhere and always a productive human principle, then we should expect to see some self-modification of the universalist claims maintained on behalf of empirical rationality" (4), Ross writes. But to my mind the second part of this sentence has no logical connection with the first. "Empirical rationality" (if by this he means the rationality that guides empirical investigations) has no obvious connection with the disposition of the fruits of that rationality or the values that yield such fruits. But then, Ross is a prolific rather than a precise thinker. He also complains about the undemocratic character of science, insofar as it fails to allow local constituents to have a say in the development of scientific knowledge (would he approve of creation science being taught in the schools?), and he pretends, here as elsewhere[4] to great sympathy for "alternative forms of rationality," such as New Age spiritualism and alternative medicines, but in a privileged "have" like him this is a familiar form of seemingly cost-free egalitarianism, a highbrow slumming, an inverted snobbery. If he were unfortunate enough to eat some half-baked chicken laced with salmonella, would he race away for a grand consult with Deepak Chopra or would he hie himself off to the emergency room? In a pinch, even an Andrew Ross is constructed (socially or otherwise) on a solid foundation of good sense. (I say this while acknowledging the value to science of citizen input and attention to alternative medicine.) And the recurring democratic-egalitarian pretensions of Ross and others in this volume, with their Promethean fantasies of bestowing speech upon the inarticulate huddled masses, seem more like the narcissism of bemused intellectuals seeking allies than a realistic program of social amelioration. Everyone not born yesterday knows that the first utterance of newly empowered proletarians to fatuous ego-inflated intellectuals is inevitably, "Go screw yourselves."

Ross's final verdict is that, piqued by reduced government funding, scientists have begun a backlash against the bad-mouthings by the science-studies left for fear of losing their wonted perquisites. This may or may not be completely false, but it is far from being a convincing exculpation of the high jinks of certain practitioners of science studies. And in referring to the "shrill tone" of Levitt and Gross, Ross belies the experience of my own prose-attuned and music-oriented ear that has been listening for most of a lifetime to how people say the things they say. If Gross and Levitt are "shrill," what would Ross have to say about Sandra Harding, whose raving essay opens this Ross-authorized collection?

"It is ironic," she begins, "that the major criticism of the new social studies of science and technology from the antidemocratic right in fact provides yet more evidence for the value of these science studies" (15). For me, "antidemocratic right" did not bode well for the level-headedness or credibility of this essay, especially when goofily reiterated in "the antidemocratic right's recent clarion calls for the citizenry to join in stamping out feminism" (17), which reads like a parody from "Doonesbury." Nor was I heartened by "Democracy-advancing social movements . . . have argued that the natural and social sciences we have are in important respects incapable of producing the kinds of knowledge that are needed for sustainable human life in sustainable environments under democratic conditions" (15). ("Democracy" has really hit the fan around here.) Harding, incredibly enough, is a professor of philosophy at the University of Delaware, which doesn't speak well for the current state of precise thinking amongst people who nowadays can pass as philosophers. Her first two footnotes defy credulity: "I use *antidemocratic right* and *democracy-advancing* movements or tendencies in a somewhat simplistic way throughout this discussion," surely the understatement of the year. And the second note offers yet another modification of her intemperate off-the-wall philosophizing: "Local knowledge systems . . . are by no means always more accurate and effective than modern scientific knowledge, but sometimes they are" (24). And sometimes professors of philosophy are hard to distinguish from idiots (but not always)! Why say stupid things in the first place if you are going to take them back in footnotes?

Writing like Harding's needs to be kept in mind when Stanley Aronowitz calls Alan Sokal "undereducated," when Andrew Ross calls Levitt and Gross "shrill," and when George Levine refers to Levitt and Gross's "hysteria" or calls their critique of science studies "unintelligent." But almost everything can be forgiven, even stupidity or mendacity, if you are a member of the club and play the right game of identity politics. After all, critics like Aronowitz, Ross, and Levine are apt to regard education as Althusserian state-supported ideologies, canons as repressive patriarchal power ploys, and the idea of "what every American should know" as a lowbrow reactionary enterprise suited only to excommunicated literary scholars like E. D. Hirsch. What on earth, then, can they be driving at by using such "exclusive," "elitist" conceptions as "undereducated" and "unintelligent"? If education is just a state-supported power grab, true freedom would consist of being as "undereducated" as possible. Let the prize be awarded to Harding, not Sokal.

In reality, however, Harding gradually comes to her senses as her essay moves along. Once the obligatory pieties have been gotten through, a little space still remains to say a few "intelligent" things instead of frothing on like Newt Gingrich. "Different organizations of knowledge generate different illuminating representations of nature" (22), so there is not just one grand version of rationality. This turns out to be her main theme, an eminently acceptable one, however familiar as "situated knowledge" in Donna Haraway and many other sources. I doubt if even those demons, Levitt and Gross, would have any quarrels here.

Steve Fuller's essay, "Does Science Put an End to History, or History to Science?" has its strengths as well as its problems. The thesis that "science" (i.e., scientific rationality) lacks any ultimate or unitary nature is bolstered by an account of the way in which the Japanese made use of Western technology while rejecting the epistemology and the Reason-with-a-capital-R that lay behind the history of science in the West. Yet even this superior narrative is compromised by its share of the equivocation and identity politics that permeate this collection. What is one to think of Fuller's candor or "intelligence" when—after alluding to Hertz, Planck, Ostwald, and the Curies, a mostly male group—he refers to their generation as "the last to be trained as 'the complete scientist,' someone who could construct *her* [my emphasis] own theories"? (29–30). Or when he refers to "physicists, chemists, and biologists" as "her"? The "dictatorship of virtue" can't have it both ways—if the science establishment has systematically excluded women over the course of its history, how is it possible to refer to the generic scientist as "her"? Or when four out of five scientists just mentioned are male, how can one's self-respect or sense of the ridiculous allow one to allude to them as "her"? How many mutually exclusive virtues is it necessary to pack into one little radical soul? Apparently, sexually exclusive language is allowable if the exclusion is politically correct, even if most male readers will not regard themselves as being addressed when they encounter female pronouns.

Although Hilary Rose, like many of the other contributors, clouds the distinction between scientific theories and social policies, she writes from a more nuanced British feminist position, able to acknowledge that "it does not follow that because scientific claims are socially shaped they are interchangeable with myths or even stories" (71). As a sociologist, furthermore, she argues that laypeople, such as sheep farmers or people suffering from obscure illnesses, often have an expertise derived from specialized experience that can extend the knowledge of professional scientists. For closed professional ranks to ignore them would be detrimental not only to social needs but to the interests of the sciences themselves, a useful application of the often murky call of this collection for greater "democracy."

Dorothy Nelkin's elaboration of science studies' critique of scientific rationality is a somewhat unpersuasive display of wounded innocence: "To some scientists this social constructivist approach appears to be a hostile attack on science, and they are responding aggressively. Indeed, their counterattack is remarkable for its emotionalism, hostility, moral outrage, and polemical tone"(93). Nelkin badly needs a reading of the letter from twenty-four "constructivist" Shakespeare feminists to Richard Levin to see emotionalism, hostility, and moral outrage in operation among her friends. And when she complains that " 'outsiders' who study science are convenient scapegoats, and waging war is an easy way for scientists to avoid critical self-inquiry" (95), I would call her attention to the "polemical tone" of Gerald Graff and his associates in cultural critique when they address the presumption of "outsiders" like David Lehman and Robert Alter for having opinions about deconstruction. But Nelkin, like the others in this volume, is more concerned with blurring the distinction between scientific rationality (with a lowercase *r*) and the social circumstances of the sciences than with being accurate. "Rather than organizing to confront the politics of the corporate state or the growing influence of religious fundamentalists" (98), scientists have organized to defend their turf, she complains. Surprise, surprise! Like Langdon Winner, who speaks of Levitt and Gross as "malicious," Nelkin appears to think that when dogs bite back after being kicked in the face, they ought to be disposed of as vicious.

The most clear-sighted and helpful essay in this volume is probably the one by Richard Levins (not to be confused with the aforementioned Shakespeare scholar, Levin). His "Ten Propositions on Science and Antiscience" attempts, with much success, to clarify the terms of the entire debate. Most analyses of science, he reports, either emphasize its objectivity while neglecting its mistakes and misuses, "or else they use the growing awareness of the social determination of science to reject its claims to any validity. They imagine that theories are unrelated to their objects of study and are merely invented whole cloth to serve the venal goals of individual careers or class, gender, and national domination" (104). I wish some of the other contributors to this volume had had a chance to read Levins before shooting their mouths off. Levins' only puzzling remark occurs near the end: "All theories are wrong which promote, justify, or tolerate injustice" (111). Would two and two have to be five if four promoted injustice? And what is injustice, anyhow? Only today I read in the *New York Times* that new laws prohibiting female circumcision in Egypt are regarded as unjust by people who have been practicing it for centuries and want to continue that practice. How would Levins handle that one?

I pass over George Levine's essay very quickly, dispirited by the arrogance and hauteur of his misrepresentations of Levitt and Gross. Implying what several others in this collection claim outright, that political correctness is mostly imaginary (while himself demonstrating its ongoing presence), this distinguished professor is content to sully his well-earned reputation as a literary scholar by traducing a pair of science scholars who come off as more dignified and trustworthy than he. Only at the end does he manage a balanced, if gratuitous, assessment of Andrew Ross. Though why he should think Ross is more worthy of serious attention than Levitt and Gross I am unable to explain.

Stanley Aronowitz's substantial and measured essay concedes that "it is difficult to deny that science has produced impressive results: rockets do reach the moon; penicillin can treat syphilis," etc. His claim "is not that science is uninfluential, only that its discoveries themselves and its influence are not unimpeachable. The import of the new social studies of science is to have shown that none of these discoveries amounts to a steady march toward Truth" (179). Agreed. (Would anyone with a brain disagree?) Yet to refer to science as "one story among many stories" (192) is at the same time both correct and misleading: music, geography, sports are other stories that represent other areas of reality. But is science just one story among many with regard to, say, Christian Science? Evangelical religion (or even mainstream religion)? Creationism? Goddess worship? There remains the attempt, even in Aronowitz, to blur the distinction between the fruits of the sciences, the politics of the sciences, and the nature of scientific rationality (even with all its errors and messiness). This is the tendentious lint that science studies needs to pick out of its meliorist fabric and which, for now, leaves me with an insuperable problem:

Namely, that somehow, for all its pious mouthings of "democracy," the academic left has become a profoundly mendacious and totalitarian establishment, crushing all voices other than its own. It is the ugly mirror image of the radical right. I ask myself once more—into whose hands would I be willing to put myself? Andrew Ross's? Sandra Harding's? George Levine's? Bruce Robbins'? Would their protestations of democratic egalitarianism let me be myself and hold onto ideas different from theirs (as Levitt and Gross and Richard Levin so obviously would)? Or would they simply dismiss me as "an outsider," "hysterical," "malicious," "unintelligent," "undereducated" and so forth, attempting to deprive me of a voice the way Stanley Fish caballed against the National Association of Scholars at Duke or the way the party-line crazies at the University of Texas tried to crush the opponents of the politically correct freshman composition text they were about to foist upon their students? If radical academia represents the best that today's democratic egalitarianism can offer, all I can say to them is, "Shove it!"

CHAPTER FOURTEEN

O, Paglia Mia!

Not so much elegiac as apostrophic—but a little elegiac too: Oh, Camille Paglia, how long can you keep this up? Even manic Italians are mortal, Dionysian as they may happen to be. Even wellsprings of energy must run dry. Even radical intelligences, when all is said and done, remain (to use a Paglism) "chthonic," sprung from earth's double-crossing clay.

But there are actually a number of Camille Paglias, one of whom sounds like this:

> But this blaming anorexia on the media—this is Naomi's [Wolf] thing—oh *please!* Anorexia is coming out of these white families, these pushy, perfectionist white families, who all end up with their daughters at Yale. Naomi arrives in England, and "Gee, all the women Rhodes scholars have eating disorders. Gee, it must be . . . *the media!*" Maybe it's that *you* are a parent-pleasing, teacher-pleasing little kiss-ass! Maybe you're a *yuppie!* Maybe *you,* Miss Yuppie, have figured out *the system.* Isn't it interesting that Miss Naomi, the one who has succeeded in *the system,* the one who has been given the prizes by the system, she who is the princess of the system, *she's* the one who's bitchin' about it? *I'm* the one who's been poor and rejected—shouldn't *I* be the one bitching about it? *No*—because I'm a scholar, okay, and she's a twit![1]

Another one sounds like this:

> Everything great in western culture has come from the quarrel with nature. The west and not the east has seen the frightful brutality of natural process, the insult to mind in the heavy blind rolling and milling of matter. In loss of self we would find not love of God but primeval squalor. This revelation has historically fallen upon the western male, who is pulled by tidal rhythms back to the oceanic mother.

It is to his resentment of this daemonic undertow that we owe the grand constructions of our culture. Apollonianism, cold and absolute, is the west's sublime refusal. The Apollonian is a male line drawn against the dehumanizing magnitude of female nature.[2]

And yet another, like this:

The cutesy treatment of clerical dress as drag, with Tallulah Bankhead cited as an authority, sets a new low for cheap vulgarity and exposes the spiritual emptiness of academe. Even a passing familiarity with anthropology or comparative religion would have helped here. But [Marjorie] Garber's interdisciplinary skills are amateurish: one of her principal sources is Vern Bullough, a contemporary archivist and unreliable popularizer. She treats history like cake batter in a swirling Mixmaster. Romanticism, the birth of modern sexual identity, is never mentioned, even apropos of Byron.[3]

The first specimen represents pop-star Paglia, the testosterone-driven rocker *manquée*; the second exemplifies Dr. Paglia, the archetypal/psychoanalytic polymath social philosopher; and the third introduces Professor Paglia, the imperious no-nonsense scholar, scourge of shoddy scholarship and screwball scholars. Pop-star Paglia does not always come off well. There's too much weighty stuff in her head for a pop icon to convey without seeming loopier than pop icons usually do. During a five-minute interview on TV, one can see behind her eyes those tumultuous oceans of thought surging for an expression even *she* can't negotiate in the allotted sound bites. So she seems frustrated, impatient, an explosion of half inarticulate emotion, a barrage of "Ex*cuse* me!" "Okay?" "Absolutely absurd!" "Gimme a break!" "Pull-*eeze*!" And thus the comic-strip effect she produces on innocent viewers: a motormouth (as the Brits call her), a crazy. With Paglia, more is better; she needs lots of expansion time; she has an immensity to say.

This need for *Sprechensraum* was marvelously well demonstrated on November 15, 1994, in a performance Paglia gave in mobbed Mandel Hall at the University of Chicago shortly after the publication of her book *Vamps and Tramps*.[4] In anticipation of the booing and riotous behavior that sometimes take place at her appearances, placards were handed out to willing members of the audience as they arrived in the vestibule, inscribed with "RESPECT CAMILLE" and similar injunctions to silence that could be held aloft if the going got rough.

Paglia appeared onstage wearing an outfit that evoked Dracula in drag, starting off extremely mannered, very uptight, all her characteristic facial and verbal tics ticking away like mad. She seemed to be anticipating immediate antagonism and was prepared to meet it by slapping the audience's faces with a whole haberdashery of gloves. Though she told us she was a lesbian and that her favorite people were gay men, she attacked the feminists, she attacked the gays, she attacked the lesbians, she even attacked the English Department of the University of Chicago. But no antagonism was displayed. No RESPECT CAMILLEs were necessary. The audience went wild with cheers and applause in what turned out to be an orgy of love and admiration. The effect on Paglia was pronounced: she became more brilliant by the minute, speaking at a furious pace with never so much as a second's pause over the course of at least two hours. Her "okay?"s and "all right?"s were spurs to Pegasus. Her knowledge of ancient civilizations, English and European literature, American pop culture—these provided a fecund reservoir of images, metaphors, illustrations, and anecdotes, all milling so close to the surface of her consciousness that she could draw upon them instantly, as the split-second shifts of her wildest improvisations required. No prompts, no notes, no nothing! Dizzying, stunning, it left one gaga.

The pop star and the professor were nicely fused at Mandel Hall into an amalgam not often found either in the media or in academia. For all Paglia's passion for pop stars like the revolting Madonna, compared to her writings about them the stars themselves seem vacuous and boring. Chthonia may be her thing, but there's no real substitute for the brains she has in such abundance. This is not to say that Paglia is without her faults, which her enemies are quick enough to attack. But their backbiting is often little more than the customary guarding of professional turf. Although other scholars are surely entitled to criticize Paglia, their narrow purview often prevents them from acknowledging her powerful accomplishments as a synthesizing public intellectual. There is, after all, a sort of general mediating knowledge more valuable than specialist scholarship, but academics are trained to disparage it. Another tactic is to dismiss *Sexual Personae* on the grounds that Paglia's treatments of major Western figures repeat things heard before or fail to reflect the very latest scholarship. Even granting its extreme repetitiveness, monomania, and need for pruning; its unconvincing overstatements; and its often suffocating projection of mythic meanings onto literature, people, and the universe at large in the now unfashionable manner of psychoanalytic criticism; even granting all of these flaws, cavils of this type minimize the astonishing synthesis, provocative philosophical foundations, polyphonic prose, and piercing intelligence required to bring off such a comprehensive performance.

Paglia's real faults are disappointing or irritating rather than fatally compromising, and to some degree they are hard to pry loose from her strengths. In *Vamps and Tramps* as in *Sex, Art, and American Culture*, she refers incessantly to *Sexual Personae* as though it were the Parthenon or Stonehenge or some monumental artifact of civilization: "my 700-page scholarly study, *Sexual Personae*" (*V&T*, xiv); "*Sexual Personae* is a Roman omnibus, a gazetteer of points of cultural transfer"; "The two volumes of *Sexual Personae* [only the first has been published so far], with the author as Amazon epic quester, may be the longest book yet written by a woman," (*SAAC*, 119). She herself seems awed by it, a Leda slammed by a Swan of Transcendence, whose issue is somehow this remarkable book. Her megalomania is stupendous: "In the four years since I arrived on the scene (after an ill-starred career that included job problems, poverty, and the rejection of *Sexual Personae* by seven major publishers), there has been a dramatic shift in thought in America" (*V&T*, xiv). Much of her reference to her work has the defensive quality of an autodidact, constantly emphasizing its *scholarliness* as opposed to everyone else's schlock. But given her knowledge and writing skills, one can only wonder why—since her work speaks for itself. (And for all her apparent megalomania, she has a very accurate estimate of her capabilities.) Puzzling too are her rivalries and putdowns, typified by the remark she made to James Wolcott for his article about her in *Vanity Fair* (September 1992): speaking of Susan Sontag, she exclaimed, "I've been chasing that bitch for twenty-five years, and at last I've passed her."

And yet these ego trips are essential components of Paglia's larger-than-life persona, even a reflection of her unusual honesty and candor, which easily coexist with the manic and hyperbolic style. She says what she feels and believes. (Are there many other gay intellectuals who would publicly assert that AIDS is a legacy of gay promiscuity in the sixties? No one at Mandel Hall booed when she said it.) She lets everything hang out, admits her past mistakes, claims not to take large fees for her lectures, continues to earn a low salary at an obscure university (still grateful for its help when she was jobless), rejects bourgeois decorum and conventional sex roles. Even as a lesbian she's not a lesbian. She tells us over and over how uncertain she finds her sexual identity and how she disbelieves in the new-style sexual essentialism that enforces its own tyrannies (i.e., you're either gay or straight and you'd damn well better admit it). She likes men, sings paeans to virility, defends patriarchy, tells gays, whom she otherwise admires, that their public behavior is sometimes outrageous (their disruptions of services at St. Patrick's Cathedral, their throwing blood on the altar, their pretense that AIDS is not foremost a gay disease [though in the years since the initial publication of her book it has proven not to be], for her it's all disgusting). To other lesbians, whom she is apt to regard as undersexed and overly "caring," she says, "Why stop at dildoes? If penetration excites, and if receptive female genitalia are so suited to friction by penis-shaped objects, why not go on to real penises?" (*V&T*, 83). To her, political loyalties mean very little.

In 1990, with the publication of *Sexual Personae*, Paglia miraculously emerged from twenty years of obscurity into a mythic notoriety. She has transformed her past into the very substance of this myth, whose components have now been repeated so many times that they have come to seem objective facts of nature, like Mount St. Helens: her Italian Catholic origins, her rootedness in an extended family, her sexual ambiguity, her fistfight at Bennington (where she was fired as prof), her inability to get jobs or dates (Who wants a pipe bomb as a colleague? Who wants to date Mount St. Helens?); her manic sixties persona, her mentoring by Harold Bloom (her depressive Other), her riotous lectures, her battles with feminists. To think of her as a mere "person" is like thinking of the Mississippi as just a river. Her writing, of course, is suffused with a mythic mentality: "Italians invented opera. It is our way of living in and reacting to the world. In opera, emotion fills the body. Italians experience emotion in sensory terms, as if it were something eaten, drunk, or poured over the flesh. Long ago on TV, Dick Cavett said he didn't like opera; he didn't 'get it.' As I looked at his small, thin body and large, smirky, Ivy League head, I said: yes" (*SAAC*, 127).

The point is not that all of this is literally true. The generalizations are too large and sweeping to be "true." Yet it is all true enough. The question to ask is, is "not literally true" the same as "off the wall"? I don't think so. Paglia's mythic style, like Kierkegaard's or Nietzsche's, is an effective vehicle for a certain type of intellectual work, an emotionally charged transvaluation of values, a melodramatic counterrevolution against political correctness. And beyond intellectual "style," there is her wild and outrageous behavior, heroic for a professor, since she admits to and lives it as part of the whole paraphernalia of her mission—it's not a secret life like Foucault's—thereby revealing how purely conventional is the outrageousness of "radical" professor types, for whom outré dogmas are requisites, not bars, to professional advancement.

Vamps and Tramps, Paglia's newest collection, is a bargain book by any criteria. Not only does one get more than five hundred pages of the most vital and adroit writing, including material not previously published (not just lintsweepings from her cupboards, as some critics have claimed), but there are TV scripts, photographs, cartoons, and comics about Paglia, and a huge annotated bibliography of writings—also about Paglia—that appeared during the two years since her previous collection. All this for a paltry fifteen dollars. Since the annotations are obviously by Paglia herself, they are often pretty hilarious.

In one sense there is nothing really new in this book. Paglia's outlook had already been made perfectly clear on page one of *Sexual Personae*:

> Society is an artificial construction, a defense against nature's
> power . . . Civilized man conceals from himself the extent of his sub-
> ordination to nature . . . Sexuality and eroticism are the intricate in-
> tersection of nature and culture. Feminists grossly oversimplify the
> problem of sex when they reduce it to a matter of social convention:
> readjust society, eliminate sexual inequality, purify sex roles, and
> happiness and harmony will reign . . . For Sade, getting back to
> nature . . . would be to give free rein to violence and lust. I agree.
> Society is not the criminal but the force which keeps crime in
> check . . . Sex *is* power. Identity is power. In western culture, there
> are no nonexploitative relationships. (*SP*, 1–2)

But in another sense the contents of this book are as fresh as ever because of Paglia's unflagging vivacity. The new essay "No Law in the Arena" attacks once more the prevailing truism among feminists and Foucauldians that human beings are socially constructed. Reaffirming that nature runs the show and that women's lives are tied to the ineradicable (and terrifyingly gory) processes of their bodies (whose psychological manifestations are unsuccessfully repressed by the corporate WASP office life many women are taught to pursue), Paglia takes the position most abhorrent to feminists that Appollonian male ratiocination and civilization building are resistances to nature and to women (the more powerful sex) as secret agents of nature. What feminists call patriarchy is civilization itself, which protects women from raw, predatory nature (like rapists and serial childbearing) by means of laws and technology (e.g., birth control pills) devised by men. If women have any chance of minimizing the grip of nature it is through male institutions, particularly those of the West. Paglia believes that resistance to nature—particularly through the arts—is humankind's chief mission, however doomed, and that the contemporary failure to acknowledge nature's power over us has led to an arrogant Rousseauist bad-mouthing of social institutions. This can only entrap women more deeply by treating them once again as fragile Victorian vessels needing protection against the society that is actually sustaining them.

Despite her unbelief, it is always Catholicism that broods over her psyche. Her "conservatism" is not the neoconservatism of politics but the conservatism of original sin. Her pet word "chthonic" is a constant reminder that to be of the earth is to be a product of nature, which some Christian churches never liked because nature, corrupted by sin and intertwined with death, is seen as the locus of all our woe. A Christian's goal is to get out of fallen nature, to become as purely spiritual as the flesh will allow. But for Paglia, steeped in the ambivalences of an eroticized, paganized, *Italian* Catholicism, the civilized "self" fights nature not by resisting it (which is virtually impossible, in any case) but by wallowing in it and then transforming it through art and artifice (in the manner of de Sade). For Paglia, it is the "unnatural" drag queen who symbolizes spirit fighting nature's totalitarianism (by giving into it—but definitely not to the point of procreation—and then transmuting it into camp). The best that the rest of us can do, our own more genteel method of giving nature the finger, is to respect society as our strongest defense, while never forgetting that nature will claim us in the end.

Seen in this light, feminists who blame society for what they detest about maleness are dupes of nature, their *real* enemy. So Paglia aims her guns at Gloria Steinem and the National Organization of Women for their "juvenile, jeering attitude toward men and masculinity" (55). She remarks upon the irony that "the legal and media world inhabited by Steinem and her coterie is filled with bookish white-collar men who are the only ones in society who actually listen to feminist rhetoric and can be guilt-tripped into trying to obey it." Or, as she puts it more succinctly (248), these are the only male office workers who have been successfully enjoined against saying, "Hey, babe! You got great tits!" ("Sensitivity-training" and the fear of lawsuits, however, may in fact have extended their own insensitive reach further than she is willing to allow.) Though Paglia thinks society should prevent men from engaging in vicious sexual harassment, she does not believe that it should (or really could) androgenize the entire human race. Without testosterone there wouldn't *be* a human race.

In later sections of *Vamps and Tramps,* Paglia goes on to consider pornography, homosexuality, and other explosive subjects with a frankness and insight that have almost disappeared from public discussions caught in the stranglehold of political correctness. In "The Culture Wars," Paglia amplifies these issues through shorter pieces collected from an impressively wide range of periodicals. Her ability to relate both classical and contemporary figures to pop culture is her distinctive signature:

> MacKinnon and Dworkin have become a pop duo, like Mutt and Jeff, Steve and Eydie, Ron and Nancy. MacKinnon, starved and weather-beaten, is a fierce gargoyle of American Gothic. With her witchy tumbleweed hair, she resembles the batty, gritty pioneer woman played by Agnes Moorehead on *The Twilight Zone.* Or she's Nurse Diesel, the preachy secret sadist in Mel Brooks's *High Anxiety.*
>
> Dworkin is Pee-wee Herman's Large Marge, the demon trucker who keeps returning to the scene of her fatal accident. I see MacKinnon and Dworkin making a female buddy picture like *Thelma and Louise.* Their characters: Penny Wise and Pound Foolish, the puritan Gibson Girl and her fuming dybbuk, the glutton for punishment. (109–10)

These identifications, not as frivolous as they seem, are "objective correlatives," concrete instantiations of Paglia's belief that pop culture artists and artifacts are contemporary versions (personae) of perennial nature-driven types, types she delineated in demonic detail in *Sexual Personae*. In practice this means her intuitions of public utterances and their speakers reinterpret them as archetypes, destabilizing the "socially constructed" foundations on which their speakers think they are standing.

Not only does Paglia criticize feminists like MacKinnon for their failure to recognize the natural, bodily imperatives that underlie human behavior and belie doctrines of social constructivism, she is engaged almost single-handedly in trying to rescue art from its subservience to feminist politics. "We will never get great art from women if their education exposes them only to the second-rate [i.e., to minor women writers, composers, and artists rescued by the feminist agenda] and if the idea of greatness itself is denied. Greatness is not a white male trick. Every important world civilization has defined its artistic tradition in elitist terms of distinction and excellence" (115). Paglia's conception of art, however, includes not only high culture's elite creations but low culture's mass media, as well as pornography—often intermixed and far from "pure," since they all strive to transmute irreducibly "chthonic" human needs. So it figures that she sees ballet and art museums as soft porn for the educated middle classes and that she defends sexy pin-ups in blue-collar lockers as working-class forms of art.

There is too much in this book for a cursory survey to sample. Articles on Princess Di, Madonna, Elizabeth Taylor, Anita Hill, Hillary Clinton, Edward Said, the Bobbitts, D. H. Lawrence, Susan Sontag reveal Paglia's extraordinary ability to make popular culture seem chthonic, complex, and "metaphysical" even to an audience as hostile to it as I am. As a book reviewer she has few peers when it comes to accurately and distinctly conveying a book's contents in clear, animated prose. Of course, her own ideologies can sometimes lead to very questionable conclusions, as when she extols the bizarre world of bodybuilding in a review of Samuel Wilson Fussell's *Muscle*, a book that minutely describes the drug-taking, eight-meals-a-day, five-thousand-plus calorie diets, and grotesque pumpings-up that characterize the daily lives of professional bodybuilders. For Paglia, their defiance of nature is enough to warrant her approval—though in reality it is ultimately *through* nature that such tortured bodies and demented personalities are produced (but this opens another complex subject). Yet nonplussing as her devotion to drag queens, bodybuilders, and rock stars may be, and joltingly raunchy as her TV scripts and video capers can appear, she comes off as a person of integrity nonetheless, with a mission defined more by Catholic saints than by the professional bottom line.[5] (What was "God" to them is "Nature" to her, and her relation to it is strongly driven by an ambivalent mixture of hatred and love.) Her writing is heavily epigrammatic, hopelessly quotable, recalling Nietzsche's in ironic reductions that expand one's insight. Far from being a flake, she represents a rare type of sanity—but one would not necessarily want to live in the same house with her.

Although he might spin in his grave to hear it, I think that Paglia—manic, hyperbolic, plugged into deafening electronic media (she watches several TVs at loud volume while composing at the computer, unable to write in silence), frequently "talking trash," violating every convention with impunity as she speaks nonstop the unspeakable—now occupies (perhaps with Henry Louis Gates) the cultural space once inhabited by Lionel Trilling, that oblique, understated, courtly, formal, elitist public intellectual of a past whose ethos seems already remotely distant. Antithetical to Trilling as her sensibility may be, her moral seriousness is as weighty as his (she shares his nature/culture and Freudian stances), and her influence on this generation is bound to be as substantial and far-flung as anything *can* be in an era in which intellectual and moral trends turn over as rapidly as materialist consumer styles. She speaks with a voice that exactly catches the air and aroma of contemporary life, more hectic than Juvenal's, a life in which (for good or for ill) academic decorum like Trilling's has been thrown to the winds, as a corrosive capitalism turns daily life into twitchy MTV videos and democracy's freedoms threaten to undermine democracy itself, reinforcing our subservience to feral nature. In keeping with times like these, her style is indecorous, and her emphasis is more on nature—and nature-in-culture—than Trilling's. She loves the products of "spirit" as much as he but insists more concretely that spirit grows from mud and mud must have its due. The startling accomplishment of her work is its relentless exposure of how much mud underlies spirit, even in the sublimest artworks, which we are likely to experience somewhat differently because of her. But whether any descendants of Trilling can learn to love the hated mud as much as she remains an open question.

Coexisting with her role of public-intellectual-as-stand-up-comedian is her scholar's instinct to conserve what has been of value in Western culture through gestures of "veneration and respect" for spiritual warriors whose victories are nevertheless forms of defeat in a battle that can't be won. Her Italian Catholicism, her defense of literature against politics, her recognition that without men there wouldn't be any women (and vice versa), her capacious aesthetic umbrella for the arts, her respect for plebian consciousness and primitive needs—these things and more distinguish her from the MLA crowd and contribute to the fame she richly deserves.[6]

CHAPTER FIFTEEN

A Crucifix for Dracula

Wendell Berry Meets Edward O. Wilson

E dward Wilson is one of those daunting scientists who write extremely well, know several fields deeply, and have been educated in the humanities during a bygone era in which culture meant more than the tawdry pages of the Sunday *Times* Arts and Leisure section. Immersed in biology, entomology, ecology, he is nonetheless familiar with literature and the arts, philosophy and literary theory, the social sciences, and much else. His book on sociobiology, reissued in a twenty-fifth anniversary edition, caused much dissention and resistance when it first appeared but has since become naturalized as a founding text in the burgeoning field of evolutionary biology. His recent book, *Consilience: The Unity of Knowledge,* pulls together much of his earlier thinking in order to promote a new synthetical direction for all the knowledge disciplines, a bold venture that has rubbed some people the wrong way.

Starting out with an extended account of Enlightenment thinkers, Wilson remarks:

> The assumptions they made of a lawful material world, the intrinsic
> unity of knowledge, and the potential of indefinite human progress
> are the ones we still take most readily into our hearts, suffer without,
> and find maximally rewarding through intellectual advance. The
> greatest enterprise of the mind has always been and always will be
> the attempted linkage of the sciences and the humanities. The ongoing
> fragmentation of knowledge and resulting chaos in philosophy are
> not reflections of the real world but artifacts of scholarship. The
> propositions of the original Enlightenment are increasingly favored
> by objective evidence, especially from the natural sciences.[1]

Although all of these sometimes controversial claims are fleshed out over and over again throughout the rest of the book, one can infer from this passage alone the foundational mindset that undergirds Wilson's thinking: conservative in the best sense of that term, Wilson believes in truth, a real world, human mindpower, and the preeminence of the sciences. For him, the astonishing feats of the mind are not ultimately the result of metaphysical intuition, faith, grace, or Platonic reminiscence but, rather, are the hard-won achievements of a material organ—the brain—that evolved along with all other forms of matter and organic life. This is a knowledge from below, unaided by nightly visits from a Miltonic Urania.[2]

The fragmentation of this knowledge, with each specialty operating according to its own rules and worldview, militates against any sort of coherent management of the problems of mankind. The split between the sciences and the humanities prevents "a clear view of the world as it really is, not as seen through the lenses of ideologies and religious dogmas or commanded by response to immediate needs" (13). Political leaders as well as public intellectuals are trained principally in the social sciences and humanities and know next to nothing about the material bases of life as described by the sciences. Natural selection "built the brain to survive in the world and only incidentally to understand it at a depth greater than is needed to survive. *The proper task of scientists is to diagnose and correct the misalignment*" (61). Wilson recognizes the vulnerability of his confidence in the abilities of the sciences but is willing to hedge his bets in favor of humankind's best hope. "Better to steer by a lodestar than to drift across a meaningless sea" (65).

Consilience, or the jumping together of the various branches of knowledge, is a concept that Wilson derives from his overall thesis about human understanding as a product of evolution. "The central idea of the consilience world view is that all tangible phenomena, from the birth of the stars to the workings of social institutions, are based on material processes that are ultimately reducible, however long and tortuous the sequences, to the laws of physics" (266). This means that to treat human faculties as special creations from above rather than growths from below is to ignore the facts of evolutionary history and the development of species to survive in congenial environments. Wilson's characterization of the field of economics can serve as a global critique of the knowledge professions in general. Speaking of the nature of classical economic theory, he remarks: "Its models, while elegant cabinet specimens of applied mathematics, largely ignore human behavior as understood by contemporary psychology and biology. Lacking such a foundation, the conclusions often describe abstract worlds that do not exist" (290).

Unsurprisingly, with a critique like this, all sorts of cherished but unwarranted beliefs about humanity get swept into the refuse bin. If, for example, the mind is a function of the innumerable circuits of the brain, an immaterial "self" that makes "free" choices becomes an unintelligible concept if it entails "freedom from the constraints imposed by the physiochemical states of one's own body and mind" (119). These conditions "consolidate certain memories and delete others, bias connections and analogies, and reinforce the neurohormonal loops that regulate subsequent emotional response. Before the curtain is drawn and the play unfolds, the stage has already been partly set and much of the script written" (119).

Even for those of us without such knowledge of the materiality of the brain, what could the notion of "free will" ever have meant—unmotivated behavior? If motivated behavior seems constraining, unmotivated behavior would be akin to insanity and madness. Certainly, we are free to weigh alternatives and can be said to have "free will" in that sense, but we are hardly free to determine the weight that the alternatives have on our constitutions, which are the products of our biological and cultural/environmental histories. Or as Wilson puts it, "*Behavior is guided by epigenetic rules*" (193), which are rules of thumb produced under the joint influence of heredity and environment. They are predisposing rather than absolutely constraining, causing us to see rainbows in four basic colors, to avoid mating with a sibling, to speak in grammatical sentences, and to fear strangers, "but they leave open the potential generation of an immense array of cultural variations and combinations" (193). And interestingly, their influence is not always benignly survival-oriented.

Sociobiology regards social customs and behaviors as offshoots of our biological needs, as a cross between genetics and psychology, between the history of the species and the history of the individual, and Wilson, in turn, sees natural selection as being more and more influenced by social, psychological, and intellectual developments, therefore only part of a reciprocal relationship rather than a sole determinant. With the development of genetic manipulation, human beings are taking greater control over their evolutionary paths, but Wilson is not receptive to open-ended genetic tinkering to create superbeings. If, as he devotes his ecological manifesto, *The Diversity of Life,* to explaining, human beings need the rain forests, earth's microorganisms, and the variety of creatures in order to survive on this particular planet, so (I infer) the human nature that took millions of years to develop cannot just be broken into with foreign supergenes without destroying a *human* ecology that makes us the species we are. Although at times he comes off as perhaps too assured about the power of the sciences, Wilson is far from being a monomaniacal hubris-driven science tyrant out of a Hawthorne short story like "Rappacini's Daughter."

When it comes to the arts, to which Wilson is far from insensitive, he shares the principles of Frederick Turner, whose "natural classicism" finds human preferences regarding meter, symmetry, regularity, balance, and so forth to be commingled with the very blood that courses through our veins—or, as Wilson would put it, guided by the epigenetic rules that formed us during the Paleolithic period. He has little that is favorable to say about academic or postmodern uses of the arts, or about postmodern nihilism and solipsism altogether, most of which he attributes to the professional needs of the various disciplines rather than any insights into our autochthonous origins and our primal sympathies with mud (so to speak). The arts are not "solely shaped by errant genius out of historical circumstances and idiosyncratic personal experience. The roots of their inspiration date back in deep history to the genetic origins of the human brain, and are permanent" (218). But because "*Homo sapiens* is the only species to suffer psychological exile," the arts have developed to "impose order on the confusion caused by intelligence" (224–25).

As for ethics and religion, the key to their history would not be found in metaphysics or philosophy but, more likely, in physics, biology, and psychology. If, I am inclined to add, our proper nutriment has in recent years been described as the diet of hunter-gatherers and not of American cattle raisers and junk-food manufacturers, Wilson's case for the aboriginal, sociobiological roots of ethical, social, and religious practices seems equally plausible, since "it would be surprising to find that modern humans had managed to erase the old mammalian genetic programs and devise other means of distributing power" (260). As a result, even though the sciences have gradually destroyed beliefs in primitive divinities and supernatural events, the need that produced these beliefs in the first place has not disappeared. (Witness the latest surge of fundamentalism.) "Science faces in ethics and religion its most interesting and possibly humbling challenge, while religion must somehow find the way to incorporate the discoveries of science in order to retain credibility. Religion will possess strength to the extent that it codifies and puts into enduring, poetic form the highest values of humanity consistent with empirical knowledge" (265).

Wilson's book concludes with what might be considered its very best chapter—on ecology and humankind's need to come to terms with the earth. Almost everything he has to say on that subject is endorsed by Wendell Berry, whose latest book, *Life Is a Miracle: An Essay against Modern Superstition* is otherwise—alas—an almost unmitigated trashing of both Wilson and *Consilience*.

Berry's prolific output of ecologically thematic poems, fiction, and essays—such as *The Unsettling of America* and *A Continuous Harmony*—is a major body of work by a Kentucky farmer and sometime academic who loathes what academia has become over the past twenty-five years. Indeed, he loathes what many things have become under the pressure of our global economy, and much of this animosity is well justified as he develops it anew in *Life Is a Miracle*. Yet there is an insidious worm that eats away at the virtues of this book and leads me to believe that Berry's eminence as a cultural guide has peaked. A voice may cry in and for the wilderness, as Berry has excellently done for many years, but no icon has a purchase on immortal wisdom, and what constituted strengths in 1975 can very well turn out to be fatally compromising in the year 2000. This was driven home to me in June of 1999 at a talk Berry gave in Kalamazoo at the third biennial conference of the Association for the Study of Literature and Environment (ASLE). To my astonishment (since he was a venerated wise man much sought after as a speaker), when his talk was done, he was attacked by some of the young graduate students and assistant professors that comprised a good part of the audience. Among other things, his positions on abortion, religion, and tobacco farming struck them as shockingly retro. Thus do idols fall!

Berry's book is essentially an assault on what he regards as Wilson's scientific hubris, his reduction of everything to biology, physics, and chemistry, his failure to pay more than lip service to a world of spirit that somehow, Berry believes, escapes the founding in materiality that generates everything else. Berry objects to Wilson's faith in science—but not to faith in general, certainly not to his own faith, and it is this extraordinary blindness that prevents Berry from seeing that he and Wilson, who appear to be mighty opposites, are essentially very similar indeed, except for their doctrines. But Wilson has the upper hand by far, because faiths, like everything else, come and go—whether in Greek gods, geocentric universes, unquestioned virtues (e.g., chastity in women but not men), or Judeo-Christian "words of God." Not even the Boy Scouts are invulnerable.

The burden of Berry's book is that science presumes to understand life and reduces it to a machine, that it has reverence for nothing, that its confidence in the powers of understanding is unfounded and, worst of all, that it does not acknowledge the mystery that lies behind phenomena, a mystery that is somehow to be understood as our interface with the sacred. Most strenuously, he objects to the "religification and evangelizing of science" and its willingness to occupy "the place once occupied by the prophets and priests of religion."[3] As for the idea of consilience, he finds Wilson's book to be "written to confirm the popular belief that science is entirely good, that it leads to unlimited progress, and that it has (or will have) all the answers" (19). Although the early pages of Berry's book manage to keep his own religious presuppositions slightly under wraps, their force gradually becomes too great to control, and by the halfway mark they burst forth in a language of defiance against the very protocols of public intellectual discourse. Even before that point is reached, however, one wants to ask him why contemporary science should *not* assume a prophetic role, as opposed to the anachronistic views of an ancient nomadic desert population whose perspectives were shaped by their own needs no less than ours are shaped by our needs. And as for the balance between theoretical and applied science, is *that* ethical track record any worse than the record of esoteric theology vis-à-vis the churches that claim to put its principles into action? If applied science has—as Berry rightly complains—allied itself with corporate capitalism and produced extraordinary damage to the environment (putting aside for the moment the stunning amelioration of human life it has also produced), is the record of the churches any better? Does one really have moral confidence in Southern Baptists who apologize 150 years too late for slavery, or a mealy-mouthed pope who plays with the truth regarding the relationship between the Church and the Holocaust and engages in doubletalk about homosexuals? (Garry Wills has been especially devastating on this subject.)

Berry goes on to complain that "a theoretical materialism so strictly principled as Mr. Wilson's is inescapably deterministic. We and our works and acts, he holds, are determined by our genes, which are determined by the laws of biology, which are determined ultimately by the laws of physics" (26). But beyond his generalized distaste for such a view, Berry has little to add that makes the notion of free will any less obscure, any less a verbal game. Even worse, "[Wilson] understands mystery as attributable entirely to human ignorance . . . He has no ability to confront mystery (or even the unknown) as such, and therefore has learned none of the lessons that humans have always learned when they have confronted mystery as such. His book is an exercise in a sort of academic hubris" (27). But what else *is* mystery but a limited state of knowledge? Things in themselves are not mysterious—they are what they are, even if *we* don't know what they are. The worship of mystery summons up visions of B movies in which the "savages" prostrate themselves in front of some phenomenon they don't understand. Of course Berry is right that everything can't be known—there isn't world enough and time—but does that confer on the residue an aura of the sacred to justify mindless idolatry, as if what can't be known differs in its quality of being from what can be? The "can't" is in the limitations of the knower, not in any intrinsic unknowability of some putatively sacred substance.

Berry criticizes Wilson's demand for evidence for religious claims while attacking him for providing insufficient evidence for his scientific optimism. "His writing about consilience is always under the sway of conditional verbs, of protestations of faith, of 'if' and 'until' and 'likely' and 'perhaps'" (36). And then he exhibits his own hubris by telling us (in scare quotes) that religious faith doesn't require any "evidence." But why should Wilson's failure to exhibit omniscience and his acknowledgment that understanding develops over time be held against him as hubris? And why should Wilson's own "faith" be disallowed for insufficient evidence when Berry's requires no evidence at all? At least Wilson attempts to provide evidence, which partakes of the ground rules for public discourse. "Scrupulous minds," Berry goes on to tell us, "must continue to live with the old proposition that some things are not knowable" (38). And yet, in the section of *Life Is a Miracle* in which he holds forth on religious faith, he confounds the possibility of public conversation by alluding to Job's "I know that my redeemer liveth" by remarking: "This statement rests on no evidence, no proof. It is not in any respectable sense a theory. Job calls it knowledge. He 'knows' that what he says is true. A great many people who have read these verses have agreed: they too know that this is so" (97). But a position of this sort—this hurling of a crucifix in the teeth of Dracula—is a confession of bankruptcy, not a surety of knowledge. For a public intellectual, it is a complete abandonment of the responsibilities of the playing field. Berry "knows" as well as I do that Job's "know" means nothing more than "believe." And if Berry really thinks otherwise, he pretty much has closed the book on his bona fides as a spokesman for anything except a narcissistic antinomianism.

Behind this retrograde display is a genuine truth: there *are* other types of knowledge besides the scientific: I *know* how I feel, I *know* that music gives me pleasure, and so forth. There is personal subjective knowing—but that this kind of knowing exists does not go very far in refuting Wilson's central claim that we have been produced out of the materials of the earth and can never cut ourselves loose from our terrestrial underpinnings, no matter how sublime our thoughts. That nurture, in other words, *is* nature. For Berry to think there are categories of thought that don't arise from below but that are handed to him directly by "God" is, for Wilson, just another atavistic earthly survival stratagem of evolution. It would involve much less hubris if we were to give up the notion that billions of years of planetary existence and the coming and going of billions of creatures has all taken place in order to make a congeries of smug believers feel good about their puny little self-regarding souls.

And indeed, there is a good deal of self-involvement on display in this book as a whole, whose constant theme is that local communities and customs, as well as the families and occupations that characterize them, most particularly *farming and farmers,* are being destroyed by the global economy, a product of applied sciences and the multinational corporations that fund them. "One of the most significant costs of the economic destruction of farm populations is the loss of local memory, local history, and local names" (138). Berry has written movingly elsewhere about the gradual destruction of the Kentucky farmlands in which he has dwelled for many decades and their takeover by faceless and exploitive corporations that care only for the bottom line. It is impossible not to read his remarks with sympathy regarding the obliteration of an entire culture that has powerful daily knowledge of and emotional ties to the land. But the cumulative effect of repeated references to this vanishing world throughout *Life Is a Miracle* is that of a *parti pris* jeremiad against the entire contemporary world for destroying Berry's lifelong habitat. And his one and only allusion to tobacco growing can only be described as chilling: "The anti-smoking campaign, by its insistent reference to the expensiveness to government and society of death by smoking, has raised a question that it has not answered: What is the best and cheapest disease to die from, and how can the best and cheapest disease best be promoted?" (145–46). A sentiment like this one does not encourage much readerly confidence in the disinterest of Berry's overall social critique. His losses, moreover, are hardly unique: *everyone* has seen the institutions, customs, and habitat of their formative years being destroyed by contemporary corporate forces: it is the old story of social change that has been experienced since day one. In a word, it's mortality. The proper response to this may not be supine quietude—but it's not a ragbag of eternal verities either.

And yet, having said this, I concur with Berry's views on professionalism, academia, and the erosion of most deeper values by money. The professions in general, and academia and the sciences in particular, have as their ends not the welfare of humankind—or even one's locality—but the interests of the professions themselves. The sciences are too often in effect "science-technology-and-industry," and as for academia, Berry asks, "If a tree falls in the absence of a refereed journal or a foundation, does it make a sound?" (62). The modern university "enforces obedience, not to the academic ideal of learning and teaching what is true, as a community of teachers and scholars passing on to the young the knowledge of the old, but obedience rather to the industrial economic ideals of high productivity and constant innovation" (63). And worst of all, "The cult of progress and the new, along with the pressure to originate, innovate, publish, and attract students, has made the English department as nervously susceptible to fashion as a flock of teenagers" (69). These critiques—provoked by Berry's reaction to Wilson's "science"—are devastatingly apposite, but I am not sure that they differ so totally from Wilson's own point of view. For Wilson also excoriates the academic professions for their insularity, self-interest, and radical ignorance of the ecological roots of all earthly existence. But what Berry would justify as the will of God is explained by Wilson as the physico-chemical-biological nature of all existent things. And what Wilson, as a scientist, attempts to support with evidence as the real nature of things, Berry, as a true believer speaking *de haut en bas,* simply "knows," along with Job, to be the truth.

CHAPTER SIXTEEN

The New Darwinism in the Humanities

From Plato to Pinker

> It may not be too much to say that sociology and the other social sciences, as well as the humanities, are the last branches of biology waiting to be included in the Modern Synthesis.
>
> EDWARD O. WILSON, *Sociobiology* (1975)

> But the intellectual climate is showing signs of change. Ideas about human nature, while still anathema to some academics and pundits, are beginning to get a hearing. Scientists, artists, scholars in the humanities, legal theorists, and thoughtful laypeople have expressed a thirst for the new insights about the mind that have been coming out of the biological and cognitive sciences.
>
> STEVEN PINKER, *The Blank Slate* (2002)

Platonic idealism—the view that mind is more real than body—may have been an epochal contribution to the lifting of mankind a few notches above the savagery of the flesh, inspiring Christianity with the sense of a "higher" and less carnalized reality that led to the Cartesian establishment of mind as autonomous and supreme. But after twenty-five hundred years of grand, self-flattering illusions about the "spirituality" and autonomy of man's unconquerable mind, a case could be made for spirituality as another, more genteel, covert form of savagery and control, another sort of narcissistic power ploy—which of course Nietzsche had already zeroed in on a century ago when he attacked it as (to coin a phrase) the guerilla warfare of the weak, "brought on by the violent severance from [man's] animal past . . . his declaration of war against the old instincts that had hitherto been the foundation of his power, his joy, his awesomeness . . . What bestialities of idea burst from him, the moment he is prevented ever so little from being a beast of action."[1] The dark side of "spiritual autonomy," "free will," and "the ghost in the machine" is not just a matter of Catholic priests revealing that they share the drives of other men or Jerry Falwell, as Jupiter Tonans, hurling hate-filled thunderbolts in the name of "God" at everybody he happens not to like. It's more serious than all that.

What if the self-confidence of the "mental," its sense of its own transcendence, its belief that it comes from above rather than from below, turned out, as per Nietzsche, to be the greatest self-deception of all, exquisitely screwing up the psyches rather than barbarously maiming the bodies of those whom it tyrannizes (though it's also done plenty of maiming)? One would want to know what, besides Plato, Descartes, church dogma, uncompromising utopian ideologies such as Marxism and Nazism, or today's mandarin political correctness could have authorized the hubris that underwrites such confidence in the autonomy of the mental, its disconnection from a materiality that keeps dragging it back down to earth anyhow?

A humility-inducing lesson could be derived from a rapid review of the evolutionary calendar, which can hardly fail to astonish a generation for whom "classic" is apt to signify little more than the venerability of a soft drink. Although the time scheme of this calendar is subject to frequent revision, a ballpark set of figures is good enough to drive home the point.

So let us say that the Big Bang, the source of all our woe, "occurred," if that's the word for it, fifteen billion years ago and that life—a one-celled sort of nothing-very-much—didn't appear until twelve billion years later. Mammals we probably wouldn't even recognize didn't emerge until about two hundred million years ago, and it was only a mere sixty-five million years ago, after the end of the dinosaurs, that reasonably familiar-looking animals entered the scene. With primates fifty million years back and hominids only seven, we are noticing a definite speedup. Still, more than another six million years had to pass before *Homo sapiens* took over, say fifty to a hundred thousand years ago. The most shocking realization of all is that the hunter-gatherer phase of hominids lasted for millions of years until, only ten thousand years ago, practically yesterday, the advent of farming introduced the settled communities we regard as civilization, which transformed human life in every conceivable way, setting off a rapid and conscious development of what today we call the arts and sciences.

Intellectual free play, that is, the use of the brain/mind for purposes other than immediate needs, is a byproduct of Darwinian selection that results in phenomena like metaphysics and computer games, whereas evolutionary psychologists connect the human brain's startling enlargement with the challenges of day-to-day survival. When bipedalism brought primates down from the trees, more intelligence was required to make tools for terrestrial living, to escape and outwit predators, and to hunt down other animals for food. Eventually, human brains became so large that surviving fetuses began to be born before they were fully viable, with heads having reached a size that overtaxed removal from the womb. Anyone who has watched the Discovery Channel or *National Geographic* on TV has seen the young of other species walking around twenty minutes after emerging from their mothers. *Homo sapiens* requires years.

Although the amazing hominid brain took billions of years to evolve from the beginnings of life, human narcissism, both religious and secular, has tried to cut it loose, as mind, from its material origins and treat it as a magical self-sustaining faculty with few predispositions. Somehow defying the parameters of all other kinds of existence, it is seen as a supposedly passive agency that can be molded like clay by churches, academies, and civil laws despite the only-too-obvious effects produced upon it not only by its evolutionary history but by food, air, water, drugs, toxic chemicals, fatigue, moods, disease, and age. As for the evolutionary and genetic pressures on brain predispositions, the grandiose notion of "human freedom" has made that a subject almost taboo. It is increasingly the task of the "modern synthesis" (an amalgamation of Darwinian evolutionary science and post-Mendelian genetics), of evolutionary biology, evolutionary psychology, and now the new Darwinism in the humanities, to counter this dangerous and overweening trend of ascribing our longings, fantasies, and productions entirely to social imprints on a blank and somehow "free" slate instead of acknowledging their mortal and finite provenance in earth-generated flesh. Indeed, it is our very material limitations that enable us to be the creatures we are: without our perceptual constraints (to use a few examples that come to mind), movies would look like a series of still photographs, television screens and computer monitors would exhibit scannings and refreshings, not moving pictures, and the music on compact disks would suffer forty-four thousand audible interruptions per second between the digital samplings. Or as Alexander Pope put it, we'd die of a rose in aromatic pain.

The publication of Steven Pinker's *The Blank Slate: The Modern Denial of Human Nature*[2] is a felicitous event affording a rich account of the foundations underlying the Darwinian interventions in the humanities to be discussed in the second part of this essay. Exhibiting all of Pinker's characteristic virtues—a lucid, demotic, incisive prose, a wide-ranging intellect, a skillful appropriation of popular culture, affability combined with straight talk, enormous learning allied with good sense—the book is destined to alter a discourse that has been held in check by political correctness and human vanity for much too long. Its founding idea, that the mind, an abstract term for the activities of a certain kind of brain—ours—is fully embedded in its matrix and not a free-floating independent entity (in fact, no "entity" at all), is hardly a new one. Even in the humanities, though scattered and fragmentary, treatments of this theme—such as Frederick Turner's *Natural Classicism,* with its vision of esthetics as expressions of primordial biological preferences—have been around for some time. But the decisive event—for Pinker and everyone else sympathetic to his stance—was the appearance in 1992 of *The Adapted Mind: Evolutionary Psychology and the Generation of Culture,*[3] a collection of essays by diverse hands, created by Jerome H. Barkow, Leda Cosmides, and John Tooby. What has become the locus classicus of the field is the book's opening essay by Cosmides and Tooby: "The Psychological Foundations of Culture," a systematic, counterrevolutionary manifesto that established the terms and issues of subsequent discourse in this arena.

The orthodoxy that triggers revolt for Cosmides and Tooby can be represented by a remark by Émile Durkheim from 1895, a sentiment whose influence shaped the social sciences for almost a century (24–25): "'Collective representations, emotions, and tendencies are caused not by certain states of the consciousness of individuals but by the conditions in which the social group, in its totality, is placed. Such actions can, of course materialize only if the individual natures are not resistant to them; *but these individual natures are merely the indeterminate material that the social factor molds and transforms.*'" (Emphasis added by Cosmides and Tooby.) From this is generated the two most powerful themes of *The*

Adapted Mind: the "Standard Social Science Model, or SSSM" and the "blank slate":

> The Standard Social Science Model requires an impossible psychology. Results out of cognitive psychology, evolutionary biology, artificial intelligence, developmental psychology, linguistics, and philosophy converge on the same conclusion: A psychological architecture that consisted of nothing but equipotential, general-purpose, content-independent, or content-free mechanisms could not successfully perform the tasks the human mind is known to perform or solve the adaptive problems humans evolved to solve—from seeing, to learning a language, to recognizing an emotional expression, to selecting a mate, to the many disparate activities aggregated under the term "learning culture." (34)

> Although most psychologists were faintly aware that hominids lived for millions of years as hunter-gatherers or foragers, they did not realize that this had theoretical implications for their work. More to the point, however, the logic of the Standard Social Science Model informed them that humans were more or less blank slates for which no task was more natural than any other. (96–97)

The appeal of the SSSM is that it provides a rationale for social engineering and political correctness, for promulgating such egalitarian absurdities as the doctrine that there are no substantive psychological differences between the sexes, a doctrine that has finally run its course. Or as Cosmides and Tooby put it, "A program of social melioration carried out in ignorance of human complex design is something like letting a blindfolded individual loose in an operating room with a scalpel—there is likely to be more blood than healing" (40). Rhetorically asking how "it is possible for pre-linguistic children to deduce the meanings of the words they hear when they are in the process of learning their local language for the first time," they reply that infants' powers of interpretation "must be supplied by the human universal metaculture the infant or child shares with adults by virtue of their common humanity" (91), in other words, their evolved nature.

Pinker's book opens up and expands upon these issues for a general audience, a fitting sequel to his previous books, *How the Mind Works* and *The Language Instinct*. His central task is to give a fatal blow to the dying orthodoxy of the blank slate, the noble savage, and the ghost in the machine. In the introduction to the twenty-fifth anniversary republication of *Sociobiology*, E. O. Wilson, speaking of Stephen Jay Gould and Richard Lewontin, writes, "They disliked the idea, to put it mildly, that human nature could have any genetic basis at all. They championed the opposing view that the developing human brain is a tabula rasa. The only human nature, they said, is an indefinitely flexible mind. Theirs was the standard position taken by Marxists from the late 1920s forward: the ideal political economy is socialism and the tabula rasa mind of people can be fitted to it. A mind arising from a genetic human nature might not prove conformable" (vi).[4] Pinker spends a goodly portion of his book amplifying the objections to this view:

> I first had the idea of writing this book when I started a collection of astonishing claims from pundits and social critics about the malleability of the human psyche: that little boys quarrel and fight because they are encouraged to do so; that children enjoy sweets because their parents use them as a reward for eating vegetables; that teenagers get the idea to compete in looks and fashion from spelling bees and academic prizes; that men think the goal of sex is an orgasm because of the way they were socialized." (x)

Pinker describes all of these as "preposterous." Bellicosity, cravings for sweets, sexual ornamentation, and male promiscuity have been well established as mating, kinship, and survival maneuvers not only among hominids and primates but to some extent among other animals as well. Far from being socially constructed, they shape the institutions of society, and far from perverting the goodness of noble savages, they are the raw materials of unreflective animal behavior. "A thoroughly noble *anything*," Pinker reports, "is an unlikely product of natural selection, because in the competition among genes for representation in the next generation, noble guys tend to finish last" (55). Along with face recognition, aversion to incest and snakes, and language acquisition, they are members of an enormous list of cross-cultural behaviors that Pinker appends to the end of this book as "Donald E. Brown's List of Human Universals." Pinker describes the predispositions on the list as "a universal complex human nature . . . of emotions, drives, and faculties for reasoning and communicating." They are "difficult to erase or redesign from scratch, were shaped by natural selection acting over the course of human evolution, and owe some of their basic design (and some of their variation) to information in the genome" (73).

As for the ghost in the machine, better known as the "self," this presents a touchy subject indeed, since it entails the concept of free will, a notion for which Pinker has little regard, though he avoids a set piece on the subject and gets by with passim remarks. But his view is clear enough: unless you accept the idea that there is an immortal human soul injected into the human body by God at the time of birth, there is no conductor of the psychological orchestra, so to speak, just billions of neurons forming systems that feel like a self. The absence of such a conductor even as we experience changes in our psychological outlooks undermines the belief that *we* (i.e., through a controlling self) "can change what we don't like about ourselves." But, Pinker asks, "Who or what is the 'we'? If the 'we' doing the remaking are just other hunks of matter in the biological world, then any malleability of behavior we discover would be cold comfort, because we, the molders, would be biologically constrained" (28).[5] For the "self" tends to be thought of "as a control panel with gauges and levers operated by a user—the self, the soul, the ghost, the person, the 'me.' But cognitive neuroscience is showing that the self, too, is just another network of brain systems" (42).

And, I would add, even if there were a magical little homunculus running the show from inside us, unless it were self-created it would be simply another collection of *données* that "we" didn't choose. And how could anything be self-created? Can a "free" and "undetermined" blank create a richly featured and desiring self? To create anything one must have drives, needs, goals, longings, emotions, preferences—in other words, a shaped character that generates behavior. Nothing can come from nothing. It's not that we "don't have free will," it's that there's nothing actual or potential that could correspond to it. It's an unthinkable thought that reveals its emptiness as soon as you try to focus on it. In sum, we're as "free" as we need to be, since the flexibility and available options for expression are immense. Witness the myriad human cultures that populate the world. It is this infinite variety that has concealed the underlying universal human predispositions. From these varied possibilities, choices (to use the passive) *are made*—if not by a "we" then by an unconscious system that makes like a we. But as motivationless "free" blanks we'd be as inert as stones, having nothing to express. It's one thing to lament not being able to fly like birds, since there are birds that actually fly. It's something altogether else to lose sleep at night about not being "free," when nothing in the universe (except perhaps for the Big Bang) is without constraining antecedents. To exist is already to be a defined and characterized *something*. It's too late to create a self ex nihilo (which couldn't be done in any case).

Pinker devotes much of his book to dealing with the fears and objections behind resistance to a critique of this trinity of obsolete metaphysical ideas—of blank slates, noble savages, and ghosts in machines. But he also wants to be clear about the dangers of rejecting one extreme in order to embrace another: "The idea of 'biological determinism'—that genes cause behavior with 100 percent certainty—and the idea that every behavioral trait has its own gene, are obviously daft" (122). If culture does not inscribe human nature upon a blank slate, neither do genes prescribe the forms in which culture realizes the genetic drives, forms that are varied beyond reckoning.

The fears that Pinker describes stem from the supposed threats to "progressive ideals" that served as platforms for the radicals of the sixties who are now the establishment. They feared inequality, differences in intelligence, differences between the races (a word that may or may not require quotation marks, depending on your political orientation). They feared imperfectability, "a permanently wicked human nature" (159) that predisposed men to promiscuity and rape, to violence and war, to selfishness. The hysterical and distorted responses to recent books on rape and on adult-child sexuality (mainly by unreflective moralists who didn't read the books) testify to the persistence of fantasies about human drives having evolved from our origins as presumably noble savages (fantasies that, as Pinker reminds us, also have their altruistic side). As for the fear of determinism, it is just a variant of the question of free will discussed above. In reply to which, Pinker's choice of a passage from Hume, like so many of his illustrative references, is wonderfully apt: "'Either our actions are determined, in which case we are not responsible for them, or they are the result of random events, in which case we are not responsible for them'" (178). And, finally, the fear of nihilism is a fear that biological explanations of the mind "may strip our lives of meaning and purpose" (186). Pinker's chapters on these fears are so discursive and nuanced that it is impossible to do them justice here.

Pinker's examination of brain development suggests that many human problems "may come from a mismatch between the purposes for which our cognitive faculties evolved and the purposes to which we put them today" (219). What we once called the soul consists of the information-processing activity of the brain, a process that can be adapted to the contemporary world by education rather than reliance on intuition, since our intuitions are too implicated in our animal history. And the education Pinker recommends for living in our high-tech society steers us toward the sciences, toward economics and biology, and away from the classical liberal arts, an ironic twist, given Pinker's own well-stocked mind.

In a section called "Hot Buttons," Pinker dwells on politics (one of the best chapters in the book), gender, violence, children, and the arts. (Again, too many riches to outline here.) "My own view," he concludes, "is that the new sciences of human nature really do vindicate some version of the Tragic Vision and undermine the Utopian outlook that until recently dominated large segments of intellectual life" (293). Yet, despite his lack of optimism about violence, human morality, unequal heritability of intelligence, ethnocentrism, and so forth, this does not come off as a pessimistic book. His own vital character as a person militates against it.

As he moves toward the finish line, Pinker turns his attention to the arts. Unlike many public intellectuals, he does not see them as going through a period of unusual trouble. Rather, he sees them flourishing more than ever. "Art is in our nature—in the blood and in the bone, as people used to say; in the brain and in the genes, as we might say today" (404). But as he reviews conflicting theories about what art is for, he does find problems. Although one of these stems from the desire for status (in the artist as a striving for novelty, and in the audience as an instance of conspicuous consumption), his main culprits are modernism and postmodernism. Taking cues from Frederick Turner regarding preferences built into our natures over millions of years, Pinker accuses modernism and postmodernism of being "based on a false theory of human psychology, the Blank Slate." They "cling to a theory of perception that was rejected long ago: that the sense organs present the brain with a tableau of raw colors and sounds and that everything else in perceptual experience is a learned social construction," which, needless to say, modernism and post-modernism have tried to shake up and disorient. But the visual system of the brain is hardly so passive: it irresistibly organizes sense data "into surfaces, colors, motions, and three-dimensional objects. We can no more turn the system off and get immediate access to pure sensory experience than we can override our stomachs and tell them when to release their digestive enzymes." Beyond this, the visual system "colors our visual experience with universal emotions and aesthetic pleasures" (412), so that people prefer savannah landscapes, beautiful faces, consonant sounds, narrative fiction, and so on. The attempts by modernist writers and artists to "make it new," to cut the connections between biologically sanctioned forms and esthetic response, has been only a partial success, as the failure of serial music has demonstrated. "Piss Christ" and "Tilted Arc," to name a few against-the-grain visual artifacts that come to mind, did not enchant their viewers, however self-satisfied their creators seem to have been.

Although Pinker enthusiastically commends a wide range of modernism's products, he is not happy with its disdain of "beauty" and its desire to frustrate our in-built nostalgia for the mud from which we spring. Moreover, the need to succeed in a market-driven society has encouraged artists to push things very far for their shock, media, and commercial value. Pinker has a warm spot for the primal directness of "middlebrow realistic fiction" because, as he believes, there is no necessary connection between the pretensions of elite high art and moral enlightenment. Quoting George Steiner to the effect that the Nazis could listen to Schubert in the morning and gas Jews in the afternoon, he is less impressed with the ethical claims of radical artists than with the unconscious psychobiological nourishment provided by more or less archetypical art forms. "The dominant theories of elite art and criticism in the twentieth century grew out of a militant denial of human nature. One legacy is ugly, baffling, and insulting art. The other is pretentious and unintelligible scholarship" (416).

I can already hear voices attacking Pinker as a philistine, but I believe they would be wrong. Pinker and E. O. Wilson are virtuoso science thinkers who have mastered the basics of contemporary humanistic culture. To accuse them of not speaking with the more subtle and complex voices of critics and theoreticians from inside the humanities would be unfair—they *aren't* insiders. They speak as superintelligent polymath outsiders, and they do a pretty good job of it. As Paul Gross and Norman Levitt kept telling us in *Higher Superstition: The Academic Left and Its Quarrels with Science*,[6] humanists in general are totally ignorant about the sciences, and their facile references to Einstein and Heisenberg make scientists laugh. Pinker and Wilson do a much more impressive job with the humanities than any humanist I know has been able to do with the sciences. They practice the consilience they recommend to others. While valuing their insights, we don't have to accept their esthetic judgments as the last word, since the matter of "beauty" in the arts is complex. We know that late Beethoven, late Wagner, Mahler, Stravinsky, Picasso, some of James Joyce and T. S. Eliot, etc. were at first regarded as "ugly" and now are so naturalized as to present few problems. What hasn't been assimilated—*Finnegans Wake, Moses und Aron*—may be the sort of artifacts that affirm Pinker's judgment.

As he concludes his overview, Pinker remarks: "Within the academy, a growing number of mavericks are looking to evolutionary psychology and cognitive science in an effort to reestablish human nature at the center of any understanding of the arts" (417). It is unnecessary to reproduce his list of luminaries here because I will turn to several of them in the second part of this account.

Back to Nature, Again

Between the year 1997, when *How the Mind Works* was published, and 2002, the year of *The Blank Slate*, Steven Pinker's treatment of art seems to have undergone a certain amount of refinement. In 1997, far from seeing the arts as "adaptive," in the Darwinian sense of conducive to fitness for survival and reproduction, Pinker described music and fiction as "cheesecake" for the mind that provided a sensual thrill like the feel of fat and sugar on the taste buds. With a view such as this, there wasn't much difference between the psychological impact of Bach's *St. Matthew Passion* and pornography off the Web. Pinker made things even worse by adding, "Compared with language, vision, social reasoning, and physical know-how, music could vanish from our species and the rest of our lifestyle would be virtually unchanged. Music appears to be a pure pleasure technology, a cocktail of recreational drugs that we ingest through the ear to stimulate a mass of pleasure circuits at once."[7]

Whether the passage of time has caused him to reconsider or whether harsh critics such as Joseph Carroll[8] have had a chastening effect, five years later in *The Blank Slate* Pinker remarks, "Whether art is an adaptation or a by-product or a mixture of the two, it is deeply rooted in our mental faculties"(405). In other words, our response to art is a component of human nature and, even if he still considers it a pleasure-technology or a status-seeking feat, Pinker now seems to see it as more deeply connected with being human. "Organisms get pleasure from things that promoted the fitness of their ancestors" (405), he writes, and he mentions food, sex, children, and know-how as well as visual and auditory pleasure. Not quite "adaptive" but serious nonetheless. If he has not already done so, I figure it is only a matter of time before he abandons the implausible view that nobody would profoundly miss music if it were simply to disappear. The number of totally music-insensitive people I have met during a lifetime would not use up the fingers of one hand.

Joseph Carroll, an English professor at the University of Missouri who can plausibly be regarded as the leading thinker among Darwinian humanists, has recently produced a brief overview of developments in this new field. He writes:

> In the past decade or so, a small but rapidly growing band of literary scholars, theorists, and critics has been working to integrate literary study with Darwinian social science. These scholars can be identified as the members of a distinct school in the sense that they share a certain broad set of basic ideas. They all take "the adapted mind" as an organizing principle, and their work is thus continuous with that of the "adaptationist program" in the social sciences. Adaptationist thinking is grounded in Darwinian conceptions of human nature. Adaptationists believe that all organisms have evolved through an adaptive process of natural selection . . . They argue that the human mind and the human motivational and behavioral systems display complex functional structure, and they make it their concern to identify the constituent elements of an evolved human nature: a universal, species-typical array of behavioral and cognitive characteristics . . . genetically constrained . . . and mediated through . . . neurological and hormonal systems that directly regulate perception, thought, and feeling . . . They are convinced that through adaptationist thinking they can more adequately understand what literature is, what its func-

tions are, and how it works—what it represents, what causes people to produce it and consume it, and why it takes the forms it does.[9]

Carroll's magnum opus, *Evolution and Literary Theory*,[10] is a powerful polemic against the post-structuralist dogmas known as textualism and indeterminacy as well as their leading exponents, Derrida, Foucault, and their many disciples. Textualism is the belief that what claims to be knowledge of a world, including the "rhetoric" of science, is only knowledge of a text, and that the attempt to make contact with a reality outside of texts is doomed by one's inability to produce anything beyond another text or rhetorical strategy. Indeterminacy, which follows from the logic of textualism, refers to the supposed impossibility of arriving at truth when all you can hope for is to produce more conflicting or self-contradictory texts disconnected from any independently existing world. In such a universe of discourse, one opinion is as good as another since none has foundations any stronger than the claims offered by the rhetorical cheering squads of each, thus leaving everything "indeterminate." The anti-poststructuralist stance of Carroll's book is a counterpart to Cosmides and Tooby's assault on the Standard Social Science Model, which sees almost everything human as a product of culture, minimally grounded in the evolved physicality of all existent things. In Carroll's case, his repudiation of the post-structuralists addresses their similar belief that everything is ultimately mental, the product of the self-enclosed human mind cut off from any constraining reality (such as "human nature" or a world). Carroll reviews in erudite detail all of the major post-structuralist theorists and, as far as I can judge, reduces them to a pile of shreds.

The positive core of Carroll's book consists of his accounts of Darwinian adaptationism and his view that "the subject matter of literature is human experience," which "is continuous with that of physics and chemistry" but which has, however, "cognitive properties that emerge only at levels of organization higher than those with which physics and chemistry are concerned, and it is these higher levels that are the appropriate subject matter of literature" (104–5). This human world is not *only* the product of culture and rhetoric, the actions of which no Darwinian would deny, but it is principally driven by the three billion years involved in the making of the human brain and is thus generated from the ground up rather than from the heavens down. "Consider," Carroll writes, "that the vast bulk of fiction consists in personal interactions constituted primarily by combinations of motives involving mating strategies, family dynamics, and social strategies devoted to seeking status and forming coalitions" (79). Among humans, this basic behavior is complicated by the peculiar human proclivity for creating elaborate cognitive models of the world and our activity in that world. For Carroll, artistic representation is a natural extension of an adaptive human capacity for creating cognitive models. In other words, "All formal literary structures are prosthetic developments of evolved cognitive structures that serve adaptive functions."[11] In still another essay, Carroll examines in concrete detail the ways in which sex, nurturing, kinship, and a multitude of evolutionary adaptations instantiate themselves in novels by Jane Austen, Charlotte Brontë, Thomas Hardy, Arnold Bennett, and Willa Cather.[12] And in one of his most brilliant essays he sums things up like this:

> I would argue that the primary purpose of literature is to represent the subjective quality of experience. In opposition to the post-Kantian notion that cognitive and linguistic categories are autonomous forms that constitute their own objects, I maintain, in company with Karl Popper, Konrad Lorenz, Tooby and Cosmides, John Bowlby, and other evolutionary theorists, that cognitive and linguistic categories have evolved in adaptive relation to the environment. They correspond to the world not because they "construct" the world in accordance with their own autonomous, internal principles but because their internal principles have evolved as a means of comprehending an actual world that exists independently of the categories.[13]

Although Darwin had a massive impact on a wide range of disciplines shortly after the appearance of *The Origin of Species* in 1859, his influence waned during the first half of the twentieth century. The resurgence of Darwinism after World War II did not really begin to transform the social sciences and humanities until, perhaps, E. O. Wilson's explosive conclusion to *Sociobiology* appeared in 1975. (That its final chapter now seems entirely unsurprising is a tribute to the extent of its naturalization over the course of twenty-five years.) And by the beginning of the nineties, the writings of Cosmides and Tooby had their own startling impact, which continues even today. What seems particularly to have generated the humanistic turn was the increasingly poisonous effect of post-structuralism in its brushing aside of the material foundations of existence along with a human nature derived therefrom and its insistence that almost everything is "constructed" by an autonomous intellect as channeled by society. (I take this up in chapter 23.) Carroll's uncompromising polemic against textualism and indeterminacy in his 1995 book seems to have produced an extremely strong humanistic influence, though even before this landmark work, Frederick Crews had made his own highly critical remarks in a brief preface to *After Poststructuralism: Interdisciplinarity and Literary Theory,* a collection of oppositional essays.[14] And even before him, in 1992, Ellen Dissanayake combated these orthodoxies in *Homo Aestheticus* (see below). More such attacks against post-structuralism followed, most notably Robert Storey's caustic dismissal of post-structuralist delusions of grandeur in the "Pugnacious Preface" to *Mimesis and the Human Animal: On the Biogenetic Foundations of Literary Representation,*[15] with an avowed indebtedness to Carroll.

Two collections of essays from the past few years provide a sense of the way in which this movement has been developing. The first, published in 1999, *Biopoetics: Evolutionary Explorations in the Arts,*[16] was assembled by Brett Cooke and Frederick Turner. "The evidence is steadily mounting," the editors remark in their introduction, "that if we wish to understand our profound and long-standing impulse to create and enjoy art we are well advised to attend to our evolutionary heritage . . . Even if art is for art's sake, it follows that we seriously consider what *that* purpose means in Darwinian terms. Not for nothing, we assume, as have many before us, is art found in every society, living or dead" (3–4). Thus the origins and rationale for the production and consumption of art are represented here by a wide, if uneven, range of essays, all of which have some connection with Darwinian adaptation and its physical and cultural consequences. Among them, the editors have collected into a mini-anthology E. O. Wilson's passim remarks on art (some very marginal) from several of his pioneering books, and Cooke has written a commentary upon them. Another contributor traces the generation of esthetic emotion to shamanistic ecstasy biochemically produced by toxic herbs or mechanically induced by drumming, chanting, fasting, or pain, all sharing aspects of sexual arousal. Yet another defines art in its most primitive manifestations as "color and/or form used by humans in order to modify an object, body, or message solely to attract attention . . . to make objects more noticeable" (265).[17]

Cooke himself, a scholar in Russian literature, provides one of the collection's few concrete readings of a literary work in adaptationist terms, examining how the treatment of women as property in Pushkin's "The Snowstorm" reflects epigenetic (i.e., the superimposition of culture upon genes) patterns of social behavior. Although these patterns are transmitted by society, the actors involved have little if any awareness of the evolutionary mechanisms that are expressed by their society's (and their own) enactment of conventions. For example, Cooke gives us the generally accepted Darwinian description of the radically different sexual behavior of males and females in most cultures:

> With gendered species, the great differential between the reproductive
> investment made by the two sexes in their offspring influences differ-
> ences in their behavior. The female generally has much less reproduc-
> tive potential than the male, and she invests significantly more time
> and energy in each offspring. The male usually makes little investment
> and, theoretically, has a vast reproductive potential. It then follows
> that the female will carefully select her mate, so as to optimize her
> limited reproduction. Male of most species may . . . try to be as
> promiscuous as possible so as to have more offspring. Some of these
> differing strategies are expressed in human behavior, such as the com-
> mon age differential between husbands and wives. (183)

Many of the underlying drives behind reproduction and nurturing may seem to be "common sense" or "logical," but evolutionists find their pervasiveness across cultures to be more than just a funny coincidence. Of course, it is possible for people "to buck the often obsolete trends of biological adaptation, but they usually will pay an emotional price for doing so" (186–87), given the lingering power of atavisms. Cooke applies these and other forces that operated during the long Pleistocene period in which we were formed to account for the essential twists and turns of the marital action in Pushkin's story—and he is pretty convincing.

Thus far, however, the number of *esthetic* evaluations of works of art from a Darwinian perspective has been small, and it is hard to say how fruitful such an approach will turn out to be. There is always the danger of forcing a variety of artifacts through a critical grinder that makes them all come out looking like the same dust. Though the range of Freudian and Marxian criticism has been great, once certain basic formulae had been applied again and again, there was an increasing tedium and self-parody involved, eliding the most distinctive aspects of artworks while distorting their character. So far, Darwinian approaches have tended to be more historical, anthropological, psychological, biological, and sociological than esthetic, so Darwinian art criticism is still in its earliest phase.

Of course it is not possible to reduce complex artworks to total conformity with any scientific paradigm, and at least one of this volume's contributors, Nancy Easterlin, has established a role as an adversarial Darwinian who tries to demonstrate ways in which culture and artworks go against the Pleistocene drives that to some degree have misfitted us for contemporary life (as Pinker insisted in *The Blank Slate,* although he regarded this going-against as more deleterious and frustrating than Easterlin does). Thus she takes the contrarian position that "works that are considered valuable and timeless are not those in which normative cognitive patterns are most closely reproduced" (243). Unlike Pinker, she is not ready to write off postmodern literary techniques and, to some degree, sees them as playing themselves off against the adaptationist norms that generate our unwitting everyday predilections.

The second collection of Darwinian essays (and there are a number of others), edited by Easterlin herself, was a special issue of *Philosophy and Literature,* a symposium on evolution and literature.[18] In it, Michelle Sugiyama writes on one of the most recurring themes in Darwinian literary study, the function of narrative: "An understanding of why and how humans create and consume narrative requires an understanding of (1) features of ancestral environments and (2) features of the mind that made the emergence of this phenomenon possible" (233). Tracing the origins of narrative far back into human prehistory, she reports on the view of anthropologists and psychologists that ritual, art, and narrative "may be conceptualized as means of exchanging information relevant to the pursuit of fitness in local habitats [during the Pleistocene]" (238). Moreover, the same themes pervade narratives worldwide, "social relations (e.g., kinship, marriage, sex, social status, morality, interpersonal conflict, deception), animal behavior and characteristics, plants, geography, weather, and the cosmos" (242). And coming much closer to home than the Pleistocene, many of these themes were already traced by Joseph Carroll in his examination of Victorian novels mentioned above. In this collection, however, Carroll (who appears in both) interconnects literature not only with evolution but with ecology as well in "The Ecology of Victorian Fiction."

> No organism can be understood except in its interactive relations
> with its total environment. An organism is never an isolated thing. By
> definition and in brute reality the world that an organism inhabits is
> part of that organism. The organism carries that world embedded and
> moulded [sic] into every inmost fold of its physiology, its anatomy,
> and its psyche . . . The felt quality of experience within a natural
> world is one of those fundamental conditions of experience. It should
> also be one of the fundamental categories of literary analysis. (302)

This joint consideration of Darwinian adaptationism and ecology has, in fact, produced the discipline of behavioral ecology. One can see how its insights might have great bearing on the creation and interpretation of literary works, given the role of place not only in nature writing but in poetry and fiction as well.

Although a Darwinism newly infused with insights from cognitive neurosciences is spreading rapidly, humanist academia so far remains a bastion of doctrinaire resistance, now that the formerly young post-structuralists are in control of English and history departments (not to mention the social sciences). The political correctness that forms the bedrock of their fundamentalism depends for its authority on the belief that people are mostly blank slates almost entirely fleshed out by culture. This belief implies that just about anything can be changed if culture so dictates. And it has been doing a lot of dictating—to a human nature that is not always very obliging. The Darwinians are seen by this opposing camp as conservatives, since their belief that the core of our being has been given rather than chosen seems restrictive and limiting, even though this human nature is expressible in infinite ways that result in individuals who are far from identical.

Culture, of course, retains great force no matter what ontology is assumed as operative: any woman living in the year 1800 in England who happened "by nature" to be athletic had little chance of satisfying athletic yearnings in a culture that forced women into a domesticity underwritten by God. Such a woman living then would have been prime material for psychiatry, a misfit neurotic who at that time could only turn to priests who reinforced the neurosis. Today, such a woman would be regarded as a model of health and would be welcomed into the world of women's sports, no psychiatrist needed. This phenomenal (in the philosophic sense) expression of the genes as culture is now being elaborated by yet another Darwin-related discipline, that of cultural biology, whose empirical investigations of brain growth reveal that both individual choices and cultural practices alter the actual physical components of hominid brains, which remain open to development throughout a lifetime (but can never be cut loose from "human nature").[19] It is only a matter of time before even humanist academia will be forced to admit that the doctrinaire truth of a truth-doubting post-structuralism is on its last legs.

I have saved Ellen Dissanayake for last because her work is the most difficult to characterize. Just before *Lingua Franca* folded at the end of 2001, Caleb Crain wrote a long account of her that began with the following summary paragraph:

> Suppose there were a person who saw, before almost anyone else, that the most important concept in modern biology could be applied to the arts. Suppose, however, that this person studied biology only as an undergraduate, never took a class in anthropology, and never received a Ph.D. Suppose, in fact, that she were a homemaker for a dozen years and then spent fifteen years in the Third World, where it was difficult for her to gain access to the research libraries and social networks that most professors take for granted. Nevertheless, over the past two decades—with no more institutional support than a few years of adjunct teaching, several grants, and a couple of visiting professorships—she has managed to publish three books setting forth her ideas. And today a new field of study has sprung up where she pioneered. Suppose, in addition, that some people think a scholarly framework based on her insights will displace much of current aesthetic theory—that future generations will understand literature and the arts as she does, thereby reconciling the humanities to the science of human nature.[20]

This heterogeneous, offbeat life is deeply relevant to Dissanayake's independent thinking and research, since she falls in neither with the orthodoxies of academic departments nor the preferred themes of the cognitive sciences, starting out with a broader experience of felt life, of the affect of behavior, than most theorists whose information depends largely on books. A Darwinian adaptationist, she connects also with human ethology, sociobiology, evolutionary psychology, psycholinguistics, neuroscience, ethnomusicology, biopoetics, developmental psychology, and much else, and her chief interest, esthetics, takes account of a wider range of human behavior than the traditional approaches.

A passage from her 1992 book, *Homo Aestheticus,* could well serve as starting point in an account of her work. Writing about the "scriptocentric" bias of modern life, she remarks:

> It seems more accurate to view thought and experience as occurring behind or beneath spoken words, as being something that saying helps to adumbrate and communicate and that writing (or rewriting) falsifies to the extent that it turns the natural products of mentation— fluid, layered, dense, episodic, too deep and rich for words—into something unnaturally hard-edged, linear, precise, and refined. We "think" like logicians primarily on (and because of) paper. If we assume that thought and experience are made wholly of language it is only because, as twentieth-century hyperliterates, we read and write reality more than we live it. (218–19)

If writing has been around for only six thousand years, and if people perform such complex activities as driving cars and playing the piano with minimal conceptualizing or attention, there's a great deal of cognition going on before the mind gets around to the discursive orderliness of speech, let alone writing. Or to put it more extremely, there's another life going on beneath the life we think we are living. And perhaps that other life is the really real one even if, or because, it can't be expressed in words.

Expression not in words is the starting point of Dissanayake's biological conception of where art comes from. In the punningly titled "Aesthetic Incunabula," both the cradle of esthetics and the cradle of an infant, Dissanayake presents her foundational theme of baby talk as the primordial expression of the arts (developed further in a series of articles and in her most recent book, *Art and Intimacy*). "Babies in every culture show the same or similar cognitive abilities and preferences."[21] The interactive baby talk in the mother-child relationship may use words, which of course the infant cannot understand at all, but it is not the words as meanings that produce the interaction; rather it is the words delivered as a form of music/poetry/dance performance, a primal esthetic experience for both mother and baby, a duet, as Dissanayake calls it, fostering emotional connection. Examining in detail a transcription of a mother's baby talk to her infant, Dissanayake reveals that beyond the infant's inborn capacity for face recognition, preference for humans, responsiveness to colors and sounds, and the adult's unpremeditated musicality of utterance to the baby, the foundations of the basic ingredients of artworks are being established:

> I suggest that what artists do in all media can be summarized as deliberately performing the operations that occur instinctively during a ritualized behavior: they simplify or formalize, repeat (sometimes with variation), exaggerate, and elaborate in both space and time for the purpose of attracting attention and provoking and manipulating emotional response. "Artification," like ritualization, attracts attention and shapes and manipulates emotion. Just as infants recognize, attend to, and respond to regularization and simplification, repetition, exaggeration, and elaboration in vocal-visual-gestural modalities when interacting with adults, so do adults attend to and respond to these features as presented to them aurally, visually, and kinaesthetically in the various arts. (343)

What Dissanayake calls "artification" here, she elsewhere characterizes as "making special." And what she consistently means by "art" is rarely elite high art of the West so much as a type of behavior. "By calling art a behavior, one also suggests that in the evolution of the species, art-inclined individuals, those who possessed this behavior, survived better than those who did not."[22] Her sense of art as "making special" was heightened by years in countries such as Sri Lanka and Papua New Guinea where customs and rituals were not as heavily overlaid by the Industrial Revolution's transformations of contemporary life in the West. Beyond ancient cave drawings, ornamentations on stone tools and handles, and the production of artifacts more beautiful than utility demanded, she calls to our attention that "each of the arts can be viewed as ordinary behavior made special (or extra-ordinary)." This is easy to see in dance, poetry, and song, which share the salient features of play and ritual, forms of exaggerated stylization of ordinary behavior. To illustrate one instance, "In song, the prosodic (intonational and emotional) aspects of everyday language—the ups and downs of pitch, pauses and rests, stresses or accents, crescendos and diminuendos of dynamics, accelerandos and rallentandos of tempo—are exaggerated . . . patterned, repeated, varied, and so forth—made special."[23] There is more here than a rapid survey can convey, but the force of her argument and the particularity of her evidence grow on you as you read a book like *Homo Aestheticus*.

"Back to Nature, Again" is, of course, sheer irony. You can't return to something you can't leave. Siamese twins, although they may not be an ideally viable life form, are as "natural" as you and I, produced by the same "laws" of chemistry, biology, and physics. There aren't any other laws. All of "us" who survive are "mutations" who have been turned into members of a species because of the serendipity of "our" adaptability. I envision a cartoon in which a group of chimps, our closest cousins, behold the first *Homo sapiens* and exclaim, "WOW! Like weird, man!" The view that we are not, in some respect, "weird" but that everything else is—as they all strive to evolve into paragons like us—is simply human arrogance and blindness. All life forms are the most natural of freaks. And our own particular freakishness is the raw material of the arts and humanities. Because they are so aware of all this, the Darwinians strike me as more "religious" than conventional religions, lacking the narcissism and hubris that can for a moment suppose that fifteen billion years of the universe and quintillions of creatures born and dead—millions at this very moment crawling all over my exterior and interior, without whom I wouldn't even exist—were produced in order to immortalize my "transcendent" little soul. (Does the universe really need my soul around forever? Do I need it?) Everything is "nature," produced from the finite materials of our planet and shaped by an aimless history with no favorites. Culture is just nature in artful and elaborate drag. In reminding us of our origins, in connecting ourselves and our arts to our biological development instead of to the heavens, the Darwinians, for me at any rate, are engaged in a long overdue hubris-crunching mission of natural piety.

CHAPTER SEVENTEEN

Ecocriticism's Big Bang

L IKE MOLIÈRE'S M. JOURDAIN speaking prose without knowing it, classic writers were unwittingly doing ecocriticism for centuries before
the genre burst forth onto the academic scene in the early 1990s. From
Virgil's *Georgics* to John Clare to Thoreau to Rachel Carson, sensitive
people had actually noticed that they were living on and from the primal
mud of Earth. Nevertheless, after many years of slow gestation, a meeting of the Western Literature Association in 1991—followed by "The
Greening of Literary Studies," a Modern Language Association (MLA)
special session in December of that same year—issued in an explicit new
discipline, a new professional organization (the Association for the Study
of Literature and Environment, known as ASLE), a new journal (*Interdisciplinary Studies in Literature and Environment,* known as *ISLE*), and
in 1996 a new canonical text, *The Ecocriticism Reader: Landmarks in
Literary Ecology,* produced by Cheryll Glotfelty and me. Ecocriticism's
early years brought together contemporary writers about nature, admiring critics of classic nature writers, and academics interested in, and con-
sumed by, the growing problems of air pollution and environmental degradation. In the decade-plus that has intervened since the birth of ASLE, the
ecocritical net has been cast over wider and wider territory to include the
ecology of cities, environmental racism, environmental law, capitalism
and colonial exploitation, and much more.

Although the cultural studies that took over the humanist academy
during the last quarter of the twentieth century have slowly begun to recognize ecocriticism, the multicultural/social-constructionist postmodern
ethos that generated them has been almost blind to the sciences upon which
any knowledge of the earth and its life depends. Ecocriticism, meanwhile,
has gradually been moving into a new and more comprehensive phase
that transcends this deficiency and acknowledges the explanatory power
of evolutionary biology and evolutionary psychology. Nonetheless, like

much study in the humanities over the past few decades, ecocriticism had early on been enabled by two fictions that have now been exhausted, one about the body and the other about the self/mind/person, aka "the soul." The first of these had to do with the "environment." The ecological movements of the past thirty years have been sustained by a distinction between the person and the environment that is wholly factitious. In this scenario, human beings live in but are semi-independent of an environment that they are harming with pollution, toxics, erosion, water usage, etc.—a dualism in which the mind, soul, or spirit retains an august autonomy derived from God or some sort of numinous stand-in, and entailing an immaculate conception in which the mind (as a "blank slate") was assumed not to have been violated by anything so gross as a body (or as Richard Dawkins has termed it, a "survival machine"). In reality, however, there is not and never has been such a thing as "the environment." Nothing "surrounds" a human being who is made of some special substance that can be distinguished from the "surroundings." There is only one congeries of earthly substance, and it comprises everything from eukaryotes to Albert Einstein.

If we could produce a high-tech time-lapse movie of the person in the environment, what would we see? A man and a woman eat food from the earth that becomes their bodies and sperm cells and eggs. A fertilized egg, fed by more plants and animals, keeps dividing, turning into specialized body parts, including a brain, that are wholly derived from the plants and animals (and the earth, sunlight, water, air, etc., that generate them). The environment is coursing through the fetus, who is made of the substances ingested by the mother. The fetus becomes a baby who becomes a person who is comprised of the plants and animals eaten by his parents and now eaten by himself. His cells, nails, hair, and skin are regularly sloughed off and replaced by newly made substance derived from earth-generated plants and animals. The person dies and decomposes back into the earth to provide food for new plants and animals to feed new parents, sperm, eggs, and fetuses.

There is no environment, only an ensemble of elements recycled through every existing thing. The environment does not wrap around the person for his regal contemplation: the person is the environment and the environment is the person. The time-lapse movie shown fast would reveal matter from the earth sweeping through the form of a person who himself sweeps back into the earth, like a wave moving across the ocean. Seen by creatures from a different time-warp, we might be indistinguishable from fruit flies. Our hominid precursors, who did not buy Krispy Kreme doughnuts or meat in plastic packages and whose genetically driven sweet tooth and need for protein meant they had to spend most of the day eating fodder as pandas do or chasing animals to acquire crucial nourishment, were more aware of this than we are. Unlike us, they literally did not know where their next meal was coming from, but when it did arrive from their hard-earned efforts, they saw very well that both they and their prey came from and returned to the same all-purpose dust. The creation myths that eventuated in later epochs reflect this primal knowledge.

As for the self/soul/spirit that seems so unmoored and amenable to culture, it is not a specially infused blank creation, like a CD-R, waiting to be formatted by any chance discourse formation or regime of truth, but a virtual projection of the brain, like the projection of a movie on a screen or on a TV. The projections look autonomous but have no independent existence and cannot initiate anything, since they are really made of thin air. They are a *trompe l'oeil*. The brain is a fantastically complex machine made of hundreds of billions of neurons that produce the sense of consciousness, sight, smell, touch, hearing, and self. But no self can be found, though just about everything else can be witnessed as brain activity by means of today's technological instruments. The desires that provoke acts of will are not chosen by a self, which cannot choose anything but which is fed by what is experienced as a stream of consciousness from inscrutable multiplex brain activity. The thoughts that move through the mind twenty-four hours a day are completely involuntary, unchosen by a "me," though my virtual "I" is moved to act (or think it is acting) on them willy-nilly. But neuroscientists now tell us that the decision to move a finger, to eat some food, to have sex, has already been produced in the brain and body a microsecond before the conscious desire arises that seems to will the activity. I, it appears, am as much a function of the environment as a bean that starts to sprout when put in moist earth or on a wet tissue.

Unless the human mind is an independent free soul injected by God into otherwise terrestrial matter, this mind is as subject to a materiality and a history as anything else. The mind may be unprecedented, amazing, astounding, plumbing the vast deeps and illimitable cosmos, but it has evolved from the same Big Bang as the cosmos and partakes of their substances, interrelations, and history. Today this whole spectacle is called Darwinian evolution or the modern synthesis, and the "human nature" it deals with is so pervasive and inclusive that Donald Brown has been able to produce an immense list of some of its characteristics, for example: esthetics, anthropomorphization, beliefs about death, body adornment, classification, collective identities, cooperation, crying, dance, empathy, figurative speech, good and bad distinguished, incest avoidance, jokes, kin groups, language, logical notion of same (and different), males more aggressive, moral sentiments, music, nouns, overestimating objectivity of thought, rituals, roles, self distinguished from other, shame, and status.[1]

The multiculturalism that dominated the humanities for the past few decades arose as a reaction to the parochial "we" that, it turned out, referred only to white, Western males and not to the human race at large. So Lionel Trilling has been taken to task for talking about the way "we" respond to Jane Austen and for a conception of human nature that was as time-bound as the psychoanalytical presuppositions of a Victorian-bred Freud. To expand this narrowness, blacks, Hispanics, Native Americans, Japanese, Sri Lankans, etc., have been taken under the wing of multiculturalism to repudiate the narrowness of "we." But if the environment is a parochial illusion, so is the seemingly broad-minded "we" of multiculturalism and diversity. Like the disparaged we of Trilling, it too is narrow and synchronic, bound to its place and time, too limited to account for very much. For the real we consists of every human being who ever lived and all the hominids and primates that preceded them. This larger diachronic we is made from the environment that comprises everything and is not just a collection of favored twenty-first-century cultures and postcolonial societies. Indeed, though it is politically correct to assert that race is a chimera and that the genetic differences between the so-called races are negligible, what tends to be overlooked, if that is true, is that the races are then ninety-nine percent the same and that the distinctive cultures that differentiate them, however worthy of study, are pretty superficial, given that we all have arrived here "out of Africa" from the consequences of the Big Bang.

If there were any doubt about the way in which today's brain and mind are tethered to a shared material past fully operative in the present, it can easily be dispelled by considering the multitudinous ways in which even at this present moment we are subject to the so-called environment. Hunger, sexual desire, fever, rage, drugs, alcohol, atmospheric pressure, air pollution, toxic substances, drought, floods, youth, age, disease—all these and more influence the way we feel and the thoughts we think at any given moment. "I" have a different psychology before food, before sex, before illness from what I have after them. At a certain point of starvation for food and sex, people will do just about anything, including cannibalism. (Think of the Donner party trapped in the snow-laden Sierras.) Afterward, they lose interest until the next round. At every moment, I am the complex production of my bodily and brain states and their immense culturally inscribed material history. A shortage of vitamin C, of protein, of trace minerals, a surfeit of refined carbohydrates, all these affect my bodily and psychological condition, my emotions, my thoughts, my point of view. Is there ever a neutral moment when I am fully an ideal healthy person (healthy according to whom?) not driven by the very particular materiality that every single second of my existence is intimately connected with? Am I free? Let's put it this way: am I unmotivated, arbitrary, the product of a vacuous, desireless blank slate? Or am I, rather, the result of my genes, my body, my country, my temporality, my family, my education, my general nurture and culture, my history, and last night's dinner—always susceptible to growth and change, however, even without an "I" to initiate it? Neo-Darwinians, after all, do not subscribe to anything as simplistic as genetic determinism, nor do they talk about nature versus nurture, whose boundaries look increasingly fluid.[2]

The decisive document in this awakening, the intellectual shot heard 'round the world, was an article by Leda Cosmides and John Tooby that appeared in 1992: "The Psychological Foundations of Culture."[3] Although it emerged from the sciences and social sciences, it is now as functionally prime for the humanities as Aristotle's *Poetics*:

> The Standard Social Science Model requires an impossible psychology. Results out of cognitive psychology, evolutionary biology, artificial intelligence, developmental psychology, linguistics, and philosophy converge on the same conclusion: A psychological architecture that consisted of nothing but equipotential, general-purpose, content-independent, or content-free mechanisms could not successfully perform the tasks the human mind is known to perform or solve

the adaptive problems humans evolved to solve—from seeing, to learning a language, to recognizing an emotional expression, to selecting a mate, to the many disparate activities aggregated under the term "learning culture."

The alternative view is that the human psychological architecture contains many evolved mechanisms that are specialized for solving evolutionarily long-enduring adaptive problems and that these mechanisms have content-specialized representational formats, procedures, cues, and so on . . . [that] tend to impose certain types of content and conceptual organization on human mental life. (34)

Although most psychologists were faintly aware that hominids lived for millions of years as hunter-gatherers or foragers, they did not realize that this had theoretical implications for their work. More to the point, however, the logic of the Standard Social Science Model informed them that humans were more or less blank slates for which no task was more natural than any other. (96–97)

As a consequence of their fatal assault on the SSSM, books on Darwin, evolutionary psychology, behavioral ecology, evolutionary biology, and so forth have been appearing more abundantly than ever. Although changes in the ethos of the humanities are now beginning to show up, they are apt to produce the startled quality of Thurber's famous "Touché" cartoon, with the slashed head of the fencer looking pretty nonplussed.

This, then, seems to be an ideal moment for the appearance of a book such as Glen A. Love's *Practical Ecocriticism: Literature, Biology, and the Environment*.[4] Love, now emeritus from the University of Oregon, has had a career in American Studies since the sixties, starting early with an ecological bent that became increasingly strong, abetted by an interest in the sciences. In his introduction he writes: "My attraction to a literal—that is, scientific—ecology and to the evolutionary biology upon which it is based has opposed a general coolness, even hostility, in the humanities toward the sciences in recent decades. Much of this hostility is an anachronistic holdover from the wholly justified reactions to the social Darwinist distortions of a century ago" (6). He gives an historical account of the growing ecocriticism movement, more or less similar to the one I have given above, and as a past president of the Western Literature Association he has witnessed the growth from inside. Although the title of his book involves a certain amount of play against the background of I. A. Richards's *Practical Criticism*, play or no play it is a good title for what follows. Not a handbook, a textbook, or a how-to book, it would serve nonetheless as an almost ideal introduction—personal or classroom—to today's ecocriticism, with its strong emphasis on science via Darwin and evolutionary biology, a book "that aims to test ideas against the workings of physical reality, to join humanistic thinking to the empirical spirit of the sciences, to apply our nominal concern for 'the environment' to the sort of work we do in the real world as teachers, scholars, and citizens of a place and a planet" (7). With its always lucid, graceful prose and its gutsiness without belligerence, it is not afraid to confront all sorts of dying shibboleths in the humanities. After three historical/theoretical chapters, Love follows through with three more exhibiting concrete treatments of Cather, Hemingway, and Howells. These exemplify a certain sort of ecocriticism in action and also reflect the academy's incipient "return to literature," which is replacing the stale iterations of yesteryear's "theory."

Love's reading has been enormously wide and deep, especially in ecocriticism and Darwinian sciences. Since my introductory remarks have already presented the foundations of his thinking, only a brief overview is needed. In his first chapter "Why Ecocriticism?" he pulls together these disciplines to characterize recent English studies "as a textbook example of anthropocentrism: divorced from nature and in denial of the biological underpinnings of our humanity and our tenuous connection to the planet"(23). This chapter describes the sorry ecological state of the planet and surveys a number of literary works that have taken cognizance of it over the years, managing at the same time to suggest the implications of evolutionary biology for both literature and life. The second chapter, on "Ecocriticism and Science," describes the science wars that reached a peak of intensity around the time of the Sokal Hoax generated by the notorious 1996 issue 46–47 of *Social Text*, which hardly needs going over again here.[5]

Love guides us through the outpouring of evolutionary books of recent decades, from the many by E. O. Wilson through Steven Pinker, Matt Ridley, Daniel Dennett, and others. For literary studies in particular, the epochal moment was Joseph Carroll's *Evolution and Literary Theory* in 1995, followed by Carroll's subsequent articles on fiction, evolution, and ecology.[6] Love remarks that "since human interaction with the biosphere is widely perceived as the defining issue of the coming century, as well as the center of ecocriticism's claim to a role in literary study, biology seems positioned for an increasingly important place in our lives" (62). If there can still be any doubt about this, two major websites alone should dispel it: *Arts and Letters Daily* (http://www.aldaily.com) and the Yahoo! group for evolutionary psychology (http://tech.groups.yahoo.com/group/evolutionary-psychology).

Love's chapter on pastoral and death recruits literary theorists and scientists to interweave connections between nature and humanity. Besides some of the already mentioned names above, he brings in Leo Marx, Stephen Jay Gould, Annette Kolodny, D. H. Lawrence, Simon Schama, Raymond Williams, Virgil and Theocritus, Lawrence Buell, Joseph Meeker, C. P. Snow, and innumerable others, with extensive reflections on E. O. Wilson's influential books. "Environmental studies," he writes, "particularly ecology, began in the life sciences and broadened to include the humanities" (63), but the need that is now more pressing is in the reverse direction. The period in which there was nothing outside the text has passed. Deconstruction's de facto revival of the New Criticism now looks stunningly inapposite—and as the Bush regime's policies for air pollution, water purity, Arctic refuges, global warming, nuclear revival, and energy consumption are added to SARS, flu, mad cow disease, and HIV in undeveloped countries, the so-called real world begins to seem very real indeed. "Man's unconquerable mind" has never seemed more vulnerable to its biochemistry.

Applying Darwinian ecocritical concepts to Willa Cather's "Tom Outland's Story" from *The Professor's House,* Love finds that it is "a particularly packed meditation on biological-cultural co-evolution . . . [Cather] looks beneath culture to its roots in human animality . . . [Her] best work demonstrates that it is not minor differences that divide humans culturally but the major similarities that unite us as a species" (105, 114, 115). When he turns to Hemingway, whom he sees as substantially influenced by Cather, Love finds a tension between a primitivism and individualism that reflects the anthropocentrism of the modern tragic hero, who glorifies a sometimes ruthless natural environment that he nonetheless destroys as part of his escape from contemporary society. In this, Love is sympathetic to Joseph Meeker's vision of comedy as an expression of Darwinian survival, as against egocentric tragedy that extols individual will even as it pulls down the natural order in acts of uncomprehending destruction.[7] With mixed feelings about *The Old Man and the Sea,* he concludes: "Hence there is more at issue in Santiago's self-doubts than Greek hubris or Christian pride. Beyond these, there is the greater folly of his assumption that the only order to the biotic world is that which his limited understanding can provide" (129).

In a long concluding essay about altruism (a major Darwinian crux) in Howells's fiction, Love concedes that Howells's evolutionism connects well with the comedy of survival but that it suffers nonetheless from the familiar exceptionalism and delusions of grandeur that raise human beings above the natural world. "The soft-Darwinian belief that mankind must distinguish itself ever more clearly from the animal world in order to achieve moral perfection does not seem to have been seriously questioned by Howells." Mark Twain, in contrast, questioned that belief "in the most caustic terms in his later works" (157). Still, Love thinks of Howells as a "realist" who ultimately sees through the utopianism of his Altrurian romances even as he exonerates the human psyche from its somatic vehicle.

All of these chapters involve critical overviews based on well-informed readings in fields that humanists generally ignore. Now and then Love overreads the ecological and evolutionary substrates of the fictions he examines, but he is mostly highly skilled and persuasive—and in the present climate of denial his counter-attempt here is almost Promethean. If the world he describes is terra incognita to so many of our colleagues, *Practical Ecocriticism* is an ideal starting point for remediation. The bibliography alone gives new meaning to "diversity."

CHAPTER EIGHTEEN

Overcoming the Oversoul

Emerson's Evolutionary Existentialism

The stern old faiths have all pulverized. 'Tis a whole population of
gentlemen and ladies out in search of religions.

EMERSON, "WORSHIP"

FLYING BACK FROM Seattle to Tucson in July 2003, rereading Emerson
for the first time in forty years before getting down to reviewing a new
book about him, looking out the window to see Mount Rainier poking
its snow-covered head through the clouds, I had a sudden vivid remembrance of things past, followed in rapid succession by a flash of insight, a
Eureka! moment. The remembrance, like a clip from an old newsreel, re-
played a scene from my almost weekly get-togethers with Joyce Carol
Oates and her husband, Ray Smith, when we were all young professors
teaching at universities in Detroit in the sixties. A recurrent field of debate, which we seemed unable to shake off, had to do with Emersonian
optimism at a time when I was teaching Emerson's essays and writing
about his religious views and their relation to Kierkegaard. Joyce Smith,
not yet celebrated as Joyce Carol Oates, was unremittingly ironic and
satiric when it came to Emerson's Oversoul and similar noumena, an
irony very pronounced even then, in her twenties. Her underlying, if un-
spoken, query in the old days was, "How can you believe such claptrap?"
She was particularly skeptical about the confidence in the decency of the
"self" that Emerson's "self-reliance" depended upon. Given her vision of
the horror and depravity that underlie human existence—fearful in her
youth and increasingly savage in her later writings—she felt that all that
could be depended upon to animate the self was a kind of primal barbarity, not the cosmic, upbeat, Rousseauvian, somewhat goofy wisdom that
Emerson often seemed to convey. And as that period of my life was fetched

up for me again while the plane made its way back to Tucson, I felt, after all these years, that hers was a challenge I had to reconsider and take seriously, though I'd now rephrase it as, "How can someone as skeptical as you be conned by such fatuous optimism that human life has cosmic meaning after all?"

After that replay had come a flash of awareness that this was a very different Emerson from the one I had read long ago or, to put it another way, this was a very different me doing the reading. Forty years had produced major intellectual revolutions that changed scholarship, literary and cultural studies, and my equipment for understanding them. Since 2003 was the bicentennial anniversary of Emerson's birth, I was about to read and review Lawrence Buell's *Emerson,* which I figured (not incorrectly) would provide a global retrospective at the start of a new century. But what I could not have guessed from my sudden illuminations was that this book would turn out to be only the beginning.[1]

After a few days at home, I opened *Emerson* and read: "Instead of concentrating on a single narrative or topical strand, [this account] provides concise intensive examinations of key moments of Emerson's career and major facets of his thought" (1). But would this extensive overview touch upon the materials of my epiphany? In the event, the absence of a single driving force or theme made this book of moderate length seem extremely long. The reader felt like someone given an opulent but fractured necklace whose beads were constantly rolling out of sight. But having said this, I need to add that Buell is one of the most intelligent, learned, and sane literary historians currently in practice, with almost forty years of involvement in Emerson studies. So although there was no single driving force, there were, nonetheless, several major themes, a few of which hovered around my own preoccupations: that Emerson's recurring engagement with the individual and his deepest "self" reflects an individualism largely purged of ego; that this self is compatible with the monism that pervades Emerson's later thinking (i.e., everything is an expression of the unity of the universe); and that "he opened up the prospect of a much more profound sense of the nature, challenge, and promise of mental emancipation, whatever one's race, sex, or nation might be" (5). Given Emerson's consistent rejection of dead traditions and his avid importation of ideas from European and Asian thinkers, which he melded into his distinctive voice as America's first public intellectual (and, indeed, its first major writer), he was always ready "to stray from paths of common wisdom into trains of thought that seem offbeat, bizarre, and sometimes downright scandalous" (5).

This straying is immediately visible in even the most brief account of Emerson's career. As a Unitarian entering the ministry, he drifted into one of the usual roles for intellectuals of his time, but in 1832, at the age of twenty-nine, he gave a sermon announcing that he could no longer in conscience administer the sacraments of the Lord's Supper (i.e., Holy Communion), because he did not believe it was Jesus' intent for this to be an ongoing practice. This effectively ended his career in the church and started him on the road to public speaking, which developed into the lecturing/touring circuit then known as the Lyceum. Although his contemporaries describe him as a mild and gentle person, his addresses and essays say No! in thunder; and, given his years of close friendship with a younger Thoreau, it is more than plausible to infer an Emersonian genesis for Thoreau's "I was not born to be forced," one of the more resounding remarks in his abolitionist essay "Civil Disobedience."

Emerson's ministerial contretemps was followed in 1837 by another shocking performance, his Phi Beta Kappa address at his alma mater, Harvard, a speech known as "The American Scholar," in which he derogated intellectuals' reliance on tradition, Europe, books, formalities, and secondhand ideas instead of on creative intelligence operating upon the actual world of nature and society. "Man thinking must not be subdued by his instruments. Books are for the scholar's idle times. When he can read God directly, the hour is too precious to be wasted in other men's transcripts of their readings."[2] ("God" in Emerson never means God, however, so this subject will require further attention below. Indeed, some of his contemporaries regarded him as an atheist.) This startling performance was followed by the even greater upheaval of 1838 when he gave an address to the Harvard Divinity School that was a religious counterpart to his trashing of scholarly timorousness and convention. This time scripture, church traditions and forms, adherence to the dead letter of custom, and the solidification of historical Christianity into a rigid myth of preposterous supernaturalisms issued in warnings that truth "cannot be received at second hand." "Miracles, prophecy, poetry, the ideal life, the holy life, exist as ancient history merely; they are not in the belief, nor in the aspiration of society; but, when suggested, seem ridiculous." Or to put it otherwise, "Men have come to speak of the revelation as somewhat [i.e., something] long ago given and done, as if God were dead." But revelation, he believed, is a permanent aspect of human consciousness, not a fait accompli that takes place only once. "Emerson's god," writes Buell, "is an immanent god, an indwelling property of human personhood and physical nature, not located in some otherworldly realm" (162). After a performance such as this, Emerson was not invited back to Harvard for thirty years, by which time he was probably the most celebrated intellectual America had yet produced.

Buell discusses these matters passim, particularly in his two best chapters, one on "Emersonian Self-Reliance in Theory and Practice" and another on "Religious Radicalisms." The possibility that self-reliance was a kind of dangerous, eccentric antinomianism leading to arrogant, ignorant, and narcissistic true believers is what made Joyce Carol Oates understandably uneasy in our debates years ago. She was by no means alone in such an opinion. "Jane and Thomas Carlyle," Buell reports, "were by turns infuriated and chastened by his saintly refusal to take offense when Carlyle attacked him head-on for moral naïveté" (313). Henry James Sr., thought Emerson a babe in the woods about the problem of evil. Charles Eliot Norton alluded (in Buell's paraphrase) to the aging Emerson's "stubbornly vacuous cosmic optimism" (315). And even his good friend Thoreau could be highly critical of his evasions. But Buell makes it clear—as does a careful reading of Emerson himself—that the self in question is the deepest, most primal and impersonal "human nature," a manifestation of the monistic force generating the universe rather than private lunacy or savage animality. This reliance, writes Buell, "requires not impulsive assertion of personal will but attending to what the 'whole man' tells you" (77). Unsurprisingly, Emerson did not find examples of such flawless self-reliance in any actual persons, since what he seems to have intended was an ideal, somewhat Rousseauvian, connection with the roots of our being, uncorrupted by the hypostatizations of transient culture, a connection expressive of the universe's deepest tendencies as manifested in the quasi-mystical "now" moments of human existence—what Heidegger and Virginia Woolf were to treat as revelatory "moments of being."

Buell has a good deal to say about Emerson's importation of Asian literature into his own poems and the subsequent congeniality of Buddhism for other writers in the American canon. Quoting both Lafcadio Hearn

and William James, he points out that "belief in a god figure is not a necessary ingredient of the religious . . . Emerson and Buddhism [according
to Hearn] stand for spirituality purged of creedal detritus" (189). This
leads Buell into extensive treatments of Emerson's influence on a wide
range of thinkers and schools, from the American Pragmatists such as
James and Dewey, who saw "spiritual 'truth' as justified by its productive
value for individual lives" (221), to Friedrich Nietzsche, about whom
Buell has much to say: "The vision of a Nietzschean Emerson also opens
up the fascinating prospect of further, indirect continental percolations
working through Nietzsche to Freud, Heidegger, and Derrida" (223).
Nietzsche was in fact an extremely admiring reader, almost a disciple, of
Emerson, and the parallel passages Buell quotes from both writers make
a pretty strong case for concrete influence. As David Mikics explains it in
a new book on this subject, "Friedrich Nietzsche discovered Ralph Waldo
Emerson in the 1860s, as a schoolboy . . . became an immediate and, as
it turned out, lifelong enthusiast of the American's work [and] quickly
discovered his crucial philosophical affinity with Emerson: a dream of individual power set against what Emerson called conformity, the common
or official beliefs that surround us."[3] And beyond this, Buell traces Emersonian influence on Whitman, Santayana, Ralph Ellison, and many others.

Although the chapter on Emerson and philosophy is perhaps overly ambitious, coming off as fragmentary, too allusive, compressed, obscure,
amends are made in richly informative chapters on Asian influence as
well as on Emerson's gradual drawing away from the abstractions of religion toward the concrete, pragmatic, everyday world of society, ethics,
and politics, goaded by the antislavery movement and the Civil War. His
summary of Emerson's work as a social thinker is worth quoting: "As a
diagnostician of the challenges of doing socially significant intellectual
work in the face of social pressure and attendant self-division, Emerson
had few equals, then or ever" (287).

Lawrence Buell touched so many bases with such profound resources
that I was surprised to find my "new" Emerson mostly neglected. Clearly,
I had a job ahead of me, a review not just of one new book but of a good
deal else that would be needed to make the case for a revised Emerson.
Other recent writings would provide some help, but there were older sup-
porting materials that I needed to consider as well.

In an essay on "Emerson and the Higher Criticism," Barbara Packer provided a useful starting point: "Throughout the eighteenth century biblical critics showed an increasing willingness to turn on the biblical texts the same principles of critical analysis that had been employed in the study of classical authors."[4] The effects of attention to internal evidence, secular history, linguistics, archaeology, editorial interventions, etc., were profound. Emerson on his own would probably have discovered the revisionary scriptural analyses of Herder, Eichhorn, and Michaelis with little difficulty, but the aftermath of his older brother William's sojourn to Göttingen to study theology in 1824 was decisive. Like Waldo, William was in training to become a minister, but after a short period of immersion in the new theology "he found his faith in the tenets of revealed religion deeply shaken by the critical questions he had been learning to ask" (74). By the time he returned from Germany, he had given up the idea of the ministry—and it wasn't so long afterward that Waldo cashiered his own clerical future by telling his Unitarian congregation that he could no longer administer Holy Communion.

From an early age, Emerson seemed to be at war with the hypostatizations of tradition and culture, which represented *other people's choices* of nutrients for the soul. In his first major work, *Nature,* published as a book in 1838, we see the effects of German Idealism and Romantic subjectivity upon this indigenous mindset. Here, the seeds of his belief that each individual has "an original relation to the universe" are profusely watered, producing the question: "Why should not we have a poetry and philosophy of insight and not of tradition, and a religion *by revelation to us* [emphasis mine], and not the history of theirs?" If "every appearance in nature corresponds to some state of the mind," we have entered the familiar Wordsworthian Romantic territory in which nature is phenomena and spirit is noumena and the task of the human person is to draw his being from whatever inscrutable force produces, organizes, and infuses the phenomenal universe—an "ineffable essence which we call Spirit." Many years later, by which time this dualism had been reduced to a monism and Emerson ceased speaking about Spirit with a capital *S,* he boldly states, "So far as a man thinks, he is free," but, he adds, "nothing is more disgusting than the crowing about liberty by slaves, as most men are, and the flippant mistaking for freedom of some paper preamble like a 'Declaration of Independence,' or the statute right to vote, by those who have never dared to think or to act."[5]

If Wordsworth rejected as inauthentic what seemed to him the excessively rule-driven poetry of Pope as well as the artificial forms of
eighteenth-century polite society, turning to infants and folk culture for
both formal and narrative features of his early poetry, Emerson (defending "children, babes, and even brutes" because "their mind being whole,
their eye is as yet unconquered") rejected a whole scholarly tradition of
secondhand bookishness in his first address at Harvard, telling the potential "American scholar" that he needs an original, self-reliant, creative relation to the universe. And by the time he addresses Harvard's Divinity
School in 1838, an entire religious and theological tradition is scan-
dalously bad-mouthed by the Emerson that some called an atheist. Religious insight cannot be received at second hand, he tells his audience, nor
can the divine nature be attributed to only one or two special persons
"and denied to all the rest." Historical Christianity "dwells with noxious
exaggeration about the *person* of Jesus. The soul knows no persons."
Rather, revelation is taking place all the time in everybody who has not
been preempted by the ossifications of tradition and its terrorizing, hell-
hurling forces. "The prayers and even the dogmas of our church are . . .
wholly insulated from anything now extant in the life and business of the
people." Jesus, the apostles, scripture, dogma—by being sanctified they
have become dead myth. "None believeth in the soul of man, but only in
some man or person old and departed."

Elaine Pagels, in what amounts to a scholarly updating of the old
Higher Criticism, has presented in her writings remarkably detailed accounts of the historically contingent procedures that resulted in certain
texts and practices being selected and reified into scriptural and ecclesias-
tical canonicity while others are ignored. It's not "God," she points out,
who is making these editorial decisions, but politically interested men
who can only think the thoughts of their own time. And from new books
on Catholicism we learn how recent dogmas like papal infallibility have
entrenched themselves in the consciousness of believers as if they were ancient laws divinely dictated. Emerson was onto all this more than 150
years ago.

Beyond seeing Jesus as just a man ("the dogma of the mystic offices of
Christ being dropped and he standing on his genius as a moral teacher"
["Worship"]), scripture as human productions of their time, miracles as
metaphorical descriptions from a credulous age, revelation (popularly un-

derstood) as just "a telling of fortunes" (rather than what it really is: "a disclosure of the soul" ["The Over-Soul"]), personal immortality as a sales pitch of Christ's disciples, beyond all these Emerson sees ordinary prayer as an act of betrayal against the regularities of the universe: "Prayer looks abroad and asks for some foreign addition to come through some foreign virtue, and loses itself in endless mazes of natural and supernatural, and mediatorial and miraculous. Prayer that craves a particular commodity, anything less than all good, is vicious . . . [Bona fide prayer] is the spirit of God pronouncing his works good. But prayer as a means to effect a private end is meanness and theft. It supposes dualism and not unity in nature and consciousness. As soon as the man is at one with God, he will not beg" ("Self-Reliance"). This dualism believes in the first instance that our universe is run by immutable regular laws but then, contradictorily, believes that these laws can be overturned through prayer by the myriad whims of a vast human population while the universe continues somehow to function regularly even as cause and effect is violated. Rather, for Emerson, human beings participate in the monistic unity of creation (which he calls, for short, the Oversoul) that runs the same through planets, rocks, plants, animals, and the consciousness of humanity. This *creator spiritus* is larger and more inclusive than any particular incarnations, so that "the philosophy of six thousand years has not searched the chambers and magazines of the soul," which could never be definitively mapped in any case, given the dialectics of consciousness ("The Over-Soul").

At the very same time Emerson was articulating these thoughts in New England, Kierkegaard was expounding similar thoughts from Denmark in an attack against an historical Christianity that he found incompatible with faith. Turning Plato upside down to view "truth" as "becoming" rather than "being," Kierkegaard saw reality as wholly dialectical: "Let it be a word, a proposition, a book, a man, a fellowship, or whatever you please: as soon as it is proposed to make it serve as a limit, in such a way that the limit is not itself dialectical, we have superstition and narrowness of spirit."[6] Emerson himself could have said this, and in fact did say it many times: "This one fact the world hates; that the soul *becomes*"! For Kierkegaard, truth is subjectivity: "that which really happened (the past) is not necessarily reality . . . There is still lacking in it the criterion of truth (as inwardness) . . . *the truth for you*. That which is past is not a reality—

for me, but only my time is . . . Historic Christianity is sheer moonshine and unchristian muddleheadedness. For those true Christians who in every generation live a life contemporaneous with that of Christ have nothing whatsoever to do with Christians of the preceding generation, but all the more with their contemporary, Christ." Kierkegaard sees historical Christianity as a sequence of scriptural interpretations, new historical findings, revisionary church doctrines, etc., etc., all subject to short-lived moments of validity. As for a potential believer waiting to justify his faith by means of these temporalities, "Just two weeks before his death he looks forward to the publication of a new work, which it is hoped will throw light upon one entire side of the inquiry," but faith—depending as it does on a subjective dialectic—will never be justified by events in the world. Meanwhile, what passes for Christianity in that mundane world is really "Christendom," a kind of faux-pious aerosol spray that Christianizes ordinary secular depravity: "Swindling has remained just as in Heathendom . . . only now the swindling has taken on the predicate of 'Christian.' So now we have 'Christian' swindling."[7]

Like Emerson, Kierkegaard writes at the inception of a movement later to be identified as existentialism, a product of Romantic subjectivity that rejects hypostatization of the past in favor of the authenticity of the on-going moments of "being" that constitute "becoming," of living in the creative power of the present moment out of which *you* make an intelligible life. Religion begins to be transformed into "religious experience," playing down history, churches, and doctrines while authenticating itself as subjectivity and, except for the literalism of fundamentalism, is henceforth to be treated by "advanced" theologians as psychology. Although Emerson was willing to countenance (for heuristic purposes) the suprahistorical force of an Oversoul, a putative world-spirit that lay within and authenticated human "Being," Kierkegaard fled from an Hegelianism that violated his deepest sense of truth as subjectivity.

But Kierkegaard was deceiving himself if he really believed that one could distinguish between Christendom and "real" Christianity or that historical Christianity could be dismissed while leaving a pure, uncorrupted, ideal residue in which one could place his faith. Christianity, like every other cultural concept and institution (e.g., Wagnerism, Freudianism), came into being at a certain time, before which there was no Christianity, neither in the flesh nor as thinkable thoughts. The Christianity in which Kierkegaard claimed to have faith was merely his own selection of data points from its history-in-the-world, in other words, from vilified Christendom. Once he eliminated historical Christianity, there was nothing left for him in which to have faith, no concepts, no vocabulary, no unsullied essence. Kierkegaard's "faith in the absurd" (a purified Christianity that defied inauthentic historical orthodoxies) was even more absurd than he imagined.[8]

For Emerson, no problem of this sort had to be faced: when he gave up on historical Christianity, he knowingly ceased to be a Christian. He had left his church, spoken of Jesus as a human role model, and used biblical history and Christian dogmas simply as figures of speech, supportive exempla in his powerful rhetoric against the dead incarnations of past spirit. If being regarded as a believer required literal belief in scripture, then, in fundamentalist layman's parlance, Emerson was indeed becoming the atheist that some of his detractors claimed. Or as Nietzsche, his devoted—but darker and more pessimistic—disciple, was famously to put it, "God is dead."

This powerful existentialist strain, sweeping its way through Nietzsche, Freud, Sartre, and many others, reached its apogee (or nadir) in the writings of Heidegger. In *Being and Time,* Heidegger carried Emersonian subjectivity and self-reliance to a point of new extremity. If Emerson had rejected historical Christianity as a series of reifications that destroyed the authority of subjectivity, Heidegger went even further—rejecting not only the Catholicism of his youth but most of the Western philosophical tradition because of its conception of reality as fixed substances rather than psyches existing in time. Heidegger invented a bizarre lexicon of neologisms—Thrownness, *Dasein,* the They, the Nothing, Unconcealment—to characterize the psychodrama of being-in-the-world. The world itself, nature, society, other people, faded into the background against which *Dasein* (literally, "being there," or individual consciousness) experienced itself in time, always open to new possibilities unless it allowed history and the They (the masses of mankind) to dictate the boundaries of human consciousness. Permitting the self to be cowed by such dictation destroyed—to use a Heideggerian buzzword—authenticity.

Being and Time is thus a vast enlargement of the theme that Emerson set forth in his "American Scholar" and Divinity School addresses, the need to leave oneself open to the unconcealment of Being. As his biographer Rüdiger Safranski summed up this version of self-reliance, "What matters in Heidegger's authenticity is not primarily good or ethically correct action but the opening up of opportunities for great moments, an intensification of *Dasein* . . . Do whatever you like, but make your own decision and do not let anyone relieve you of the decision and hence the responsibility."[9] Like Kierkegaard, to whom he was greatly indebted, and Emerson (to whom the debt is unclear), Heidegger transformed philosophy and theology into an oftentimes solipsistic psychology, a psychology more extreme than that of his progenitors and, compared to Emerson's (and even Nietzsche's), peculiarly morbid, with its focus on anxiety and death.

After Heidegger's overwhelming interrogation of Being, it would seem there could be little future remaining for existentialism, but this was assuredly not the case. Psychology, psychiatry, and psychotherapy, combined with questionings of both scriptural literalness and church authority, produced virtually a century of existential theology, writings that attempted to translate difficult ideas of philosophy and psychology into a more popular and therapeutic religious language. Today, however, these manifestoes more often than not come off as maudlin and stale. Books like Paul Tillich's *The Courage to Be* (with its existential, somewhat desperate, doubletalk about God as the "ground of being," as "ultimate concern," as "the God beyond God") and Martin Buber's *I and Thou* fanned flames that eventuated in the writings of Bishop John Robinson of Woolwich (a London suburb), whose *Honest to God* launched the Death of God movement of the 1960s. More recently, the writings of emeritus Bishop John Spong of Newark, New Jersey, ask "Is There a Future for the Christian Church?" and explain *Why Christianity Must Change or Die,* a case of déjà vu all over again (and this time, Yogi Berra's malapropism is *le mot juste*). Behind all this furious revisionist activity to flog the Anglican and Episcopal churches back to life lurk the ghosts of Emerson, Kierkegaard, Nietzsche, and Heidegger.

With repeated references to Kierkegaard, Nietzsche, Tillich, Bonhoeffer, and Buber, Bishop Robinson dismisses as preposterous the traditional anthropomorphic God in the sky, a "being" who "exists" like any other in time and space, with a personality, preferences, emotions. He treats scripture, yet again, as myth and metaphor. In place of the God "out there," he adopts Tillich's language of depth, God as the ground of our being. "The question of God," he writes, "is the question *whether this depth of being is a reality or an illusion,* not whether a Being exists beyond the bright blue sky." At the point of "love to the uttermost," he continues many pages later, "we encounter *God,* the ultimate 'depth' of our being . . . This is what the New Testament means by saying that 'God was in Christ' and that 'what God was[,] the Word was.'"[10]

When I examined my shelf of religious writings from the sixties, I was reminded of the sizable industry generated by Robinson's book: *The Death of God, The Death of God Controversy, The Honest to God Debate, Radical Theology and the Death of God, The Secular Meaning of the Gospel,* and so on. Tumbling out of these volumes were typed carbons of letters I wrote to both Robinson and William Hamilton, an offbeat Episcopal theologian, as well as their replies. In a letter of three single-spaced, densely typed, vexed pages, I asked Robinson a series of blunt questions, particularly challenging his preposterous claim as to what the writers of the New Testament *really* meant. And with all of the supernaturalism removed, I suggested, there was nothing much left of revisionist Christianity that differentiated it from conventional secular morality except yet another set of neologisms. Robinson wrote a graciously evasive reply: "I just do not accept that what I am trying to say is so totally out of line with traditional Christianity as you assume," he repeats unconvincingly several times.

In "The Death of God Theologies Today," William Hamilton was less genteel:

The breakdown of the religious *a priori* means that there is no way, ontological, cultural or psychological, to locate a part of the self or a part of human experience that needs God. There is no God-shaped blank within man . . . As Protestants, we push the movement from church to world as far as it can go and become frankly worldly men. And in this world, as we have seen, there is no need for religion and no need for God. This means that we refuse to consent to that traditional interpretation of the world as a shadow-screen of unreality, masking or concealing the eternal which is the only true reality.[11]

Hamilton's outlook and my own in 1966 were just about identical liberal humanist views, with this major difference: Hamilton persisted in using traditional Christian terminology. But what difference existed between us apart from the terminology or what these words could now mean in their eviscerated condition I was unable to see. In my letter to him, I compared his use of "Jesus" to Crest toothpaste's use of "Fluoristan," "which in fact is the same fluoride present in all other similar toothpastes . . . Jesus is just your patented name for a combination of qualities found elsewhere." If I simply changed my vocabulary, I asked him, would I be turned into a Christian like him, instead of a secularist whose beliefs were almost exactly like his own?

His reply was genial. "There may be actually no form of being in the world that the radical knows that some unbeliever doesn't also know, but the radical Christian takes this historical figure [i.e., Jesus] as a model, paradigm, focus of loyalty, though wooden and lifeless forms of mimicry have to be watched." Moreover, "A Christian is a man who somehow allies himself with the Christian community in some form, whatever that means. A man is defined, in part, by his choice of comrades . . . But the fact remains that whatever our similarities might be, Jesus is the name of the difference." This struck me then and strikes me now as pretty thin, not to say desperate, as a basis for spiritual rehabilitation. The mountains have labored and produced a mouse.

But wait, as they say on TV infomercials, there's more! In the latest wave of Christian demystifications, Bishop Spong has produced "Christ and the Body of Christ: Is There a Future for the Christian Church?" a distillation of his book *Why Christianity Must Change or Die: A Bishop Speaks to Believers in Exile*.[12] The essay touches on most of the issues found in Robinson and his theological siblings from forty years prior, but in a take-no-prisoners finale, it concludes with a list of twelve theses summarizing his message and, indeed, existential revisionist theology altogether:

1. Theism, as a way of defining God[,] is dead. God can no longer be understood with credibility as a Being supernatural in power, dwelling above the sky and prepared to invade human history periodically to enforce the divine will. So most theological God-talk is today meaningless unless we find a new way to speak of God.
2. Since God can no longer be conceived in theistic terms, it becomes nonsensical to seek to understand Jesus as the incarnation of the theistic deity. So the Christology of the ages is bankrupt.
3. The Biblical story of the perfect and finished creation from which

human beings fell into sin is pre-Darwinian mythology and post-Darwinian nonsense.

4. The virgin birth, understood as literal biology, makes the divinity of Christ, as traditionally understood, impossible.

5. The miracle stories of the New Testament can no longer be interpreted in a post-Newton world as supernatural events performed by an incarnate deity.

6. The view of the cross as the sacrifice for the sins of the world is a barbaric idea based on primitive concepts of God that must be dismissed.

Resurrection, ascension, ethics inscribed in stone, prayer to change events, life after death, churchly behavior-control through guilt—all these are trashed in the remaining theses. And finally, speaking against bigotry and prejudice in number twelve, Spong concludes that "All human beings bear God's image and must be respected for what each person is." Except for the "God's image" phrase, which is incomprehensible in the context of these twelve theses (but which is a telling sign of the conflicting paradigms that plague revisionist theology), this is a powerful, clear, forthright, and courageous statement. Nevertheless, it is little more than a reinvention of the wheel. There is not much new here that we had not heard from Emerson 150 years before. We have, in a sense, come full circle, an eternal recurrence, a flogging of a horse long since dead.

Where does all this leave us today with regard to Emerson and his legacy at the bicentennial of his birth? Emerson was a religious seer who rejected historical Christianity in particular and incarnations in general but who never relinquished the prophetic, missionary persona that animated his writings. Although he is indeed a co-father of existentialism with Kierkegaard, for some reason he has not generally been acknowledged as such. Even in Robert Denoon Cumming's *Starting Point*,[13] a philosophical history of existentialism, Emerson cannot be found in the index, and the major emphasis there is on Kierkegaard and Heidegger. And how, in this undeniably existential light, is it possible for us to understand the seemingly numinous Oversoul, always hovering in the wings but, like the Holy Ghost, impossible to photograph? For a thinker who compulsively swept the Augean stables of moonshine (while providing plenty of his own), what can we make of this seeming inconsistency, this throwback into spooks (even though Emerson didn't mind being inconsistent)? What can I say to Joyce Carol Oates in defense of Emerson today?

Emerson was a prime mover behind revisionary interpretations of Christianity since the early nineteenth century. Although even earlier thinkers such as Hume, Gibbon, and Voltaire (to name the most notable) had attacked Christianity and superstition, they usually did so under the cover of self-protective irony. Emerson, however, living in a less constricting age and society, was not inclined to pull his punches, either rhetorically or in his short-lived vocation as a Unitarian minister. This is not to say that he was not agonized by risky decisions he needed to make with regard to his career as a sage. In *Understanding Emerson: "The American Scholar" and His Struggle for Self-Reliance,*[14] Kenneth S. Sacks provides an intimate portrait of Emerson's life during his most dangerous years. Fearful of alienating the audiences of his Lyceum talks, a major source of income, he was very cautious about what he said and didn't say to laymen about religion and politics. By fellow intellectuals he was sometimes criticized for indecisiveness and a failure to follow through in the real world, though by traditionalists he was attacked as a madman. But when push came to shove, Emerson plowed boldly ahead. The "American Scholar" and Divinity School talks can make a reader catch his breath even today, and his essays in general rarely fail to shock. The self-reliance he preached in lieu of conventional scholarship and historical Christianity was not the self-involvement of a besotted narcissist. On the contrary, as these things go, it was rigorously impersonal. So who or what was this self that warranted so much deference?

Emerson's views changed markedly during his lifetime as he followed new developments in the sciences, so that he was faced with the acrobatic task of reconciling his growing acceptance of materialism with his sense of the spiritual foundations of the phenomenal world. In two of his most powerful late essays, "Experience" and "Fate," Emerson makes it clear that everything derives from matter, including the mind, and there can be no "spiritual" self or "free will" in the commonly used senses of an incorporeal soul enacting unmotivated behavior, a nonsense concept. On the one hand, he remarks, "So far as a man thinks, he is free," but on the other, he takes back a lot: "If we thought men were free in the sense, that, in a single exception one fantastical will could prevail over the law of things, it were all one as if a child's hand could pull down the sun. If, in the least particular, one could derange the order of nature,—who would accept the gift of life?" Although he remained optimistic until the end, his thinking became progressively darker as his awareness of human constraints grew more cosmic. He saw human beings as completely woven into the material web of the universe. There was only one sort of substance, not two. No Cartesian dualism could infect his sense of a monosubstantial universe that embraced all things.

Yet despite his willingness to countenance so much darkness, Emerson, like his existential theological offspring described above, needed to have some kind of escape valve from "meaninglessness," because he saw the universe as a whole and human life in particular as essentially "moral," despite their radical materiality. Morality indeed had been the theme of his early book *Nature*, the unity of mind and world, and it remained at the center of his thinking until the end, despite the considerable metamorphoses induced by science. In "Fate," the escape valve was power, the human ability to countervail, to outwit the chains of causation through intelligence and creative thinking. "Intellect annuls Fate," he optimistically declared. Man "betrays his relation to what is below him—thick-skulled, small-brained, fishy, quadrumanous—quadruped ill-disguised, hardly escaped into biped, and has paid for the new powers by loss of some of the old ones. But the lightning which explodes and fashions planets, maker of planets and suns, is in him." This is the sort of marvelous stuff that makes Emerson so lovable after two centuries, even if he noted but didn't want to expatiate on *thinking* as part of the same chain of causation as everything else. "Even thought itself," he concedes, "is not above Fate: that too must act according to eternal laws," but after noting it he conveniently forgets it in order to proclaim the emancipating force of brainpower.

How, then, are we to make sense of his contradictions, his hard-boiled realism and his cloud-nine Romantic idealism? What could possibly reconcile the existential self, human "freedom," the Oversoul, and the universe, somehow joining them together into a reasonably coherent view of reality? The answer seems to be that for Emerson, as surprising as it may seem, science in general and evolution in particular (indisputably visible in the quote just above) generated the spiritual glue that held his worldview together.

In her recent definitive book, *Emerson's Life in Science: The Culture of*

Truth, Laura Dassow Walls gives us the most thorough picture available of the massive role of the sciences in Emerson's thinking, speaking, and writing. "A complete survey of Emerson's reading in science would fill many volumes . . . Goethe, Schelling, Lorenz Oken, Georges Cuvier, Jean-Baptiste Fourier, Laplace, Adolph Quetelet, Davy, Faraday, Lyell, Kerschel, Alexander von Humboldt, Agassiz, Darwin, and Tyndall, among others." Although "his most intensive reading in science occurred from 1830 to 1834, the years leading up to and following his resignation from the ministry and his first trip to Europe," he "continued to read widely in science right into the 1870s."[15] The open-endedness of scientific discovery coordinated well with his sense that reality was "becoming" rather than "being," and the ability of the sciences to bestow order on the universe reinforced his sense of reality as mind-driven, rational, moral. His visit in 1833 to the Muséum d'Histoire Naturelle in Paris provided a powerful moment, recorded in his notebook, in which he saw specimens of insects, birds, and animals artfully arranged to reveal their phylogeny, in his words "an occult relation between the very scorpions and man," or as Walls describes it, he saw "the organizing idea which had created them" (85). From the earliest essays and lectures, his writings are dominated by scientific references and figures of speech. He absorbed changes in scientific doctrines as they occurred and incorporated their lexicons into his rich and quirky prose.

Very early on, more than thirty years before Darwin's *Origin of Species* in 1859, Emerson had begun to use evolutionary language to describe the organization of life on this planet. His contemporary and friend, Moncure Conway, a minister with his own religious crises, remarks in his autobiography: "We who studied him [i.e., Emerson] were building our faith on evolution before Darwin came to prove our foundations strictly scientific."[16] And in his *Emerson at Home and Abroad,* he dwells for several pages on the Darwinian aspects of Emerson's thought: "It was perfectly clear to him that the method of nature is evolution, and it organized the basis of his every statement." Conway even recognized what we now would describe as Emerson's existentialism: "The old phrases 'Supreme Architect,' 'Almighty,' 'Providence,' had become fossil to him whose deity had become subjective."[17] Yet there is almost no comment about Darwin in all of Emerson's writings beyond *obiter dicta* about obtaining or reading a copy of *Origin.* Conway makes tantalizing allusion to a discussion he had with Emerson about Darwin when they met in Cincinnati, but what they said remains a cipher beyond his own reference to their circle as "pre-Darwinite Evolutionists." Ralph Rusk, editor of the Emerson letters, ventures a tentative footnote that begins, "Emerson, who had been for many years an interested spectator of the march of science and a student of earlier speculations on evolution, must have been deeply stirred by Darwin's great book,"[18] but Laura Dassow Walls seems convincingly on target when she dryly remarks, "When Emerson came to read Darwin, which he did in 1860, he saw nothing he had not seen before—a fact that reveals little about Darwin and a great deal about Emerson" (167). What he had already seen were the writings of Lyell, Cuvier, Lamarck, Tyndall, Robert Chambers, and a host of other writers who touched, in one way or another, on evolutionary subjects before Darwin. None of these writers posed a threat to Emerson's belief system because, as Conway put it, "Emerson held no such theism as could be affected by any scientific discovery or opinion."

The most penetrating account of Emerson's saturation with evolutionary ideas is still Joseph Warren Beach's "Emerson and Evolution" from 1934.[19] Beach traces the evolution of Emerson's ideas about evolution, from his early Coleridgean years involving a static "scale of being" with man at the top, more or less familiar from Pope's "Essay on Man," to a quasi-Darwinian assent to the "transmutation" of species, with the higher emerging from the lower. In the 1830s, Emerson accepted the notion of a not as yet *evolutionary* progress toward human beings, especially after the impact of his visit to the Paris museum: "There has been a progressive preparation for him [man] . . . the meaner creatures containing the elements of his structure . . . His limbs are only a more exquisite organization—say rather the finish—of the rudimental forms that have been sweeping the sea and creeping in the mud."[20]

Beach sees Emerson's reading of Lyell and Lamarck as planting seeds that were "bound to sprout," moving him past the "scale of being" stage and its supposition of a divine intervention that produced new species, to an increasingly evolutionary one, as in his second essay on nature: "It is a long way from granite to the oyster; farther yet to Plato and the preaching of the immortality of the soul. Yet all must come, as surely as the first atom has two sides." By 1844, when he read *Vestiges of the Natural History of Creation,* by Robert Chambers, it was, as Beach describes it, "the most plausible and comprehensive view of evolution that Emerson had ever encountered. But so familiar was this way of thought by then that it caused him neither shock nor excitement" (488). Chambers, however, rubbed Emerson the wrong way theologically: "What is so ungodly as these polite bows to God in English books? . . . Everything in this Vestiges of Creation is good, except the theology, which is civil, timid, and dull" (488). In 1854, he proclaimed that "the creation is on wheels, in transit, always passing into something else," but he added a critical caveat: "The ends of all are moral, and therefore the beginnings are such . . . everything undressing and stealing away from its old into new form, and nothing fast [i.e., locked in place] *but those invisible cords which we call laws, on which all is strung*" (490–91; emphasis mine). Moreover, the laws of nature and the laws of thought were the same, so that if their true order were to be found, the poet could "read their divine significance orderly as in a Bible" (492). If this were so, then "Why should we fear to be crushed by savage elements, we who are made up of the same elements?"

Emerson's most startling and definitive statement on this subject appeared in a lecture from 1858:

> You do not degrade man by saying, Spirit is only finer body; nor exalt him by saying Matter is phenomenal merely . . . You will observe that it makes no difference herein whether you call yourself materialist or spiritualist. If there be but one substance or reality, and that is body, and it has the quality of creating the sublime astronomy, of converting itself into brain, and geometry, and reason; if it can reason in Newton, and sing in Homer and Shakespeare, and love and serve as saints and angels, then I have no objection to transfer to body all my wonder and allegiance.[21]

By the time Darwin's *Origin* appeared in 1859, Emerson's evolutionary views had been pretty well formed.

Both Joseph Beach and Laura Dassow Walls, however, see Emersonian doctrine as falling far short of Darwinism. "The more he learns of natural history," Beach writes, "the more certain he is that it is all a projection of the mind, an expression of the inherent moral purpose of the universe which is found in the human spirit" (494). Emerson was so disposed to see the laws of nature as intrinsically ethical that he took for granted that ethical concepts were embedded in "the intellectual system of the universe. He never glimpsed the idea that *ethical concepts may be themselves the product of evolution*" (496; emphasis mine). Walls, in an essay on Emerson and Victorian science, sums him up as "a transcendental idealist, not a transcendental realist, willing from the start to concede, even celebrate, the role of the mind in making experience possible." And given the *practical* orientation of his morality, in the final analysis she characterizes him as a pragmatist who never really understood the "Darwin who deposed Providence and enthroned chance as the governing power of the universe. This was the lesson of natural selection, the engine that drove evolution and Darwin's real innovation."

Emerson saw order where Darwin saw happenstance. But Emerson's order was not the *Fiat lux!* of an external Providence that directed the universe. Zeus and his Christian avatars, after all, were dead. Rather, for him it was the internal rationality of the constituent elements of the universe itself. And herein, as Walls so aptly points out, lay Emerson's biggest misunderstanding of all: "The key here lies in Emerson's understanding of 'laws of science,' which starting in the twentieth century came to be merely descriptive rather than constitutive, but which in Emerson's day were still understood to 'govern' nature, such that they should also 'govern' us."[22] Thus no transcendent lawgiver was needed, since the raw materials of the universe were themselves legislative. The religious enterprise of the individual self was to decipher these laws through the promptings of its deepest being, an existential task that entailed the casting aside of society's anachronistic directives in order to find "the truth for me." For Emerson, the "problem," whatever it was, had been solved. But for us? What finally can we make of it all at this late date?

Rereading Emerson for the first time in four decades, I saw light bulbs flashing in ways that would have been impossible forty years earlier, before the Darwin-inspired "modern synthesis"[23] and the growth of today's cognitive sciences and evolutionary psychology. And just as it is necessary now to give a heliocentric interpretation to an older text's geocentric conclusions in order to try to understand what the author was driving at, there is no way I can now read Emerson as if the Modern Synthesis had not taken place.

Had Emerson lived his life several decades later, into the early years of the next century, when William James[24] and Nietzsche were familiar presences, the residues of his nineteenth-century German Romantic Idealism would have faded even faster than they did in his own era. Emerson's "spirit" and "spirituality" had already become less and less numinous, more and more material as he took in the writings of his vanguard contemporaries. His monism came to settle on matter, not thought, as the primal substance, and his gradual movement away from the ghostly immanence of the transcendent (whatever that oxymoron could have meant) into the somewhat less ghostly moral "law" embedded in matter follows a clear pattern. In 1836 he could write, "Idealism sees the world in God" as phenomena. But in 1837 he would tell his audience, "Out of unhandselled[25] savage nature . . . out of terrible Druids and Berserkers[26] come at last Alfred and Shakespeare."

More and more he refers to "human nature," to one universal mind that becomes less and less transcendent, to babes, idiots, and savages being closer to nature. His familiar and unfamiliar quotables trace an evolutionary development that blurs the distinction between one and all: "to believe that what is true for you in your private heart is true for all men"; "In all conversation between two persons tacit reference is made, as to a third party, to a common nature"; to "involuntary perceptions a perfect faith is due"; "all the facts of history pre-exist in the mind as laws"; and "Strong race or strong individual rests at last on natural forces, which are best in the savage, who like the beasts around him, is still in reception of the milk from the teats of Nature." By 1858, "God" was just a figure of speech, and he was saying, "Spirit is only finer body."

The *meaning for us* (if I may steal Emerson's existential thunder) is patent: the Oversoul, the immanent laws, the universal mind and what pre-exists in it, and even "God," as Emerson uses these terms, are what today we would refer to as human nature—but now understood in the terms prevailing in the post-Darwinian cognitive sciences. Put very crudely, it means that everything human comes from the biochemical stuff of which we have been made throughout our evolutionary history. Nothing comes simply from "outside" because consciousness mediates all experience— and consciousness has evolved along with everything else. Nurture is not outside. Everything experienced by a subject is ultimately immanent. In a sense that Laura Dassow Walls did not think of when she correctly remarked that "the laws of science" are descriptive rather than legislative, human ethics and spirituality really are legislated by the stuff of which we are composed. (Recall Joseph Beach's remark that Emerson "never glimpsed the idea that ethical concepts may be themselves the product of evolution.") Given creatures like us, with bodies and brains like ours, "human nature" necessarily produces the myriad arts, sciences, ethics, customs, and religions that comprise the totality of human cultures—as well as the savagery from which they arise. No adventitious spooks are required to account for this—the "laws" of biochemistry and physics are spooky enough. Emerson's bothersome "self," in which he had so much confidence, was not the narcissistic "individual," who sometimes turns out to be a savage "berserker" suited to a story by Joyce Carol Oates. The "self" for Emerson was impersonal, universal. *It was radical human nature*. Radical, as in *"of the roots."*

At the end of *The Blank Slate: The Modern Denial of Human Nature*, Steven Pinker thoughtfully appends an alphabetized list of universally found characteristics of *Homo sapiens*, taken from Donald E. Brown's *Human Universals*. The list is long, but here are a few representative samples: esthetics, anthropomorphization, beliefs about death, body adornment, classification, collective identities, cooperation, crying, dance, empathy, figurative speech, good and bad distinguished, incest avoidance, jokes, kin groups, language, logical notion of same (and different), males more aggressive, moral sentiments, music, nouns, overestimating objectivity of thought, rituals, self distinguished from Other, shame, statuses, and roles. Here are found the springs of the legislated "ethics" inscribed in the universal "self." Emerson lived in a period in which "spirit" and "transcendence" and "Oversoul" were still some of the ways of talking about these things. One hundred and fifty years later, the Oversoul is biochemistry is evolution is spirit is culture is ethics is human nature.

In recent decades the distinction between nature and nurture has gradually been eroding. Even after dismissing the foggy notion of "free will," few cognitive thinkers consider us automata driven by inflexible genes. Matt Ridley's *Nature via Nurture*, Daniel Dennett's *Freedom Evolves*, Quartz and Sejnowski's *Liars, Lovers, and Heroes*, William Calvin's *A Brain for All Seasons*, Pinker, E. O. Wilson, and the others are gradually arriving at a view of genes as devices that switch on and off depending on environmental and mental/physical states, as well as brains that physically alter on the basis of experience. Matt Ridley gives example after example in which it is impossible to call a human response genetic or cultural or experiential. Thinkers such as Peter Singer and Roderick Nash had already spoken of the widening circle of human empathy (as in the case of animal rights), which they saw as evolvements of human nature as developed by culture. Still, survival generally requires looking out for number one, so nobody expects a human nature that *prefers* "the Other" to one's individualized self, though the Darwinians speak of "inclusive fitness," the proclivity of nonreproducing people to abet the procreation of their more fit kin. Human nature, even amidst barbarity and berserkers, is inscribed with ethics after all.

Emerson's existentialism is in sharp contrast with the visions implied by his heirs, the existential theologians quoted above. Despite their eagerness to trash historical Christianity and its superstitions, in their heart of hearts they long for a renovated orthodoxy with a new and vaporous lingo to replace the old, even as they hang on to a body they have killed and eviscerated, befuddled about how they can bring it back to life. Emerson's worldview, repelled by orthodoxies, was open-ended, evolving, unspecified, rejecting all incarnations as strictly *pro tem.* Apart from his belief in what I here am calling human nature, he had no institutional doctrines to offer. He would feel quite at home with the legacy of Darwin and its recent cognitive developments. Like Darwin himself, Emerson lacked a sense of tragic finality, even as his purview continued to darken. My guess is that he found the last sentence of *Origin of Species* worthy of assent: "From so simple a beginning endless forms most beautiful and most wonderful have been, and are being, evolved."

I would also venture to guess that Joyce Carol Oates and I, now forty years down the pike, would have a meeting of the minds were we to resurrect our bygone, half-joking argumentations about that troublesome old Oversoul.[27]

May it rest in peace.

CHAPTER NINETEEN

Back to Bacteria

Richard Dawkins' Fabulous Bestiary

"F**ABULOUS**" SUGGESTS A FABLE, but Richard Dawkins' *The Ances-tor's Tale*[1]—a reverse journey of sorts from *Homo sapiens* to the primal blob—is in large part fact, in slightly smaller part inspired speculation, and in still smaller part artful fabrication. Only a master of the game of evolutionary history could have produced an opus as grandly magnum as this one. To create his journey back to the parent of us all, Dawkins has founded his six-hundred-page epic on an act of poetic license that probably causes more trouble than it's worth. Acknowledging that a retro-history of evolution back to square one could very well begin with any extant creature, he nonetheless (bowing to "human interest") chose *Homo sapiens* as his startup vehicle, while deciding to treat the journey as a pilgrimage in the style of *The Canterbury Tales*. In the persona of a Host, he picks up a "pilgrim" at each point at which a species branch reconnects (since we're going backward) to a larger branch of the evolutionary tree, a point in other words where, in retrospect, we can identify a new taxonomic lineage as having arisen. These pilgrims are actually progenitors of the new species, common ancestors Dawkins has neologized as "concestors," most of whom, at least in theory, tell a "tale," like Chaucer's pilgrims.

One wishes that this literary device worked out better than it does, since in reality there is no Host, no pilgrims, no tales and no Canterbury, just Dawkins as the grand narrator who speaks in a number of voices, not in order to imitate diverse pilgrims (who are nowhere in evidence) but to employ the rhetorical mode that his story requires at each turn. These modes range from genial, literary, knockabout informal discourse to highly technical set pieces in the specialized language of zoology. I would call this virtuoso performance an oratorio—with recitatives, stately arias, and maybe an occasional grand chorus—more like Haydn's *The Creation* than Chaucer's *Canterbury Tales*. The result is a book that is at once awe-inspiring and not quite satisfactory.

A multimodal performance such as this raises the question of what constitutes a "book," or at least a book that produces a distinct and powerful impression. The familiar Richard Dawkins, celebrated for such cultural artifacts as *The Selfish Gene* and *The Blind Watchmaker* as well as collections of essays and reviews such as *A Devil's Chaplain,* is here only fitfully in evidence, mainly in the meditative arias. The long discursive unwrappings of a single theory or insight that drives his well-known works provides them with a continuity of narrative and voice that serve as a motive force largely lacking in *The Ancestor's Tale.* There are a lot of dry (but densely informative) zoological recitatives describing the major life forms along the way, admittedly the heart of the book. The most gripping parts are the discourse-rich early pages presenting general ideas, the periodic "arias" in which Dawkins steps back from his ongoing bestiary to speculate and ruminate about the significance of its zoological particulars or to hurl political, religious, and scientific thunderbolts at his *bêtes noirs,* and the final pages in which he attempts an overview and summation. But six or eight pages on the electrical fields of platypuses are bound to fatigue the most indomitable of nonspecialists. In the course of six hundred pages, one is likely to wonder who is the intended audience.

Dawkins was justified in his supposition that starting at the beginning of reproductive life perhaps three and a half billion years ago and moving forward to the present would have given the impression of a progress toward us (an evolutionary no-no), whereas going backward avoids such an anthropocentric assumption, squashing our grandiosity by reducing us to the blobs of bacteria from which we and all other life emerged. As he puts it, "We can be very sure there really is a single concestor of all surviving life forms on this planet. The evidence is that all that have ever been examined share (exactly in most cases, almost exactly in the rest) the same genetic code; and the genetic code is too detailed, in arbitrary aspects of its complexity, to have been invented twice" (7).

How is it possible to learn so much about life forms from the distant past, many of them extinct? Dawkins offers three sources of information: archaeology, renewed relics, and triangulation. Archaeology studies bones, teeth, pots, artifacts, as well as fossils that have survived for millions of years, some unearthed by digs, others by having been compressed into formations like the Burgess Shale (the subject of Stephen Gould's *Wonderful Life*) that reveal even soft tissue. Renewed relics are accounts found in written records, such as literary works and discoveries like the Dead Sea Scrolls. But writing goes back only five thousand years, a mere blip in the record of life on earth. The real archival golden relic is DNA. Although the actual molecules of dead animals don't last very long—mostly days or years, but "for plants in permafrost, the record is about 400,000 years"—nevertheless the information in those molecules is "copied for millions, sometimes hundreds of millions of years" (21) in subsequent generations whose DNA turns out to be a record of the past, preserved like digital copies of a compact disk even when the original vehicle is destroyed. As for triangulation, the most speculative of the three techniques, Dawkins gives an optimistic report: "Even if we had no fossils, a sophisticated comparison [i.e., triangulation] of modern animals would permit a fair and plausible reconstruction of their ancestors" (23).

With these basic investigative tools explained, Dawkins sets out on his journey to the source, starting with *Homo sapiens* and regressing through forty branchings to arrive at primal bacteria. The first backward split or branching occurred about five million years ago, when our line broke off from that of chimps and bonobos, our closest relatives (which means sharing very similar genomes). Bipedality and brain enlargement provide two of the most speculative cruxes in evolutionary biology, since they are the driving forces behind the acceleration of culture and technology. Many theories to explain these cruxes derive from the renewal of interest in Darwin's *The Descent of Man* for its introduction of the concept of sexual selection, the libidinal trigger behind mating preferences, which are principally the whims of females dazzled by displays of male fitness. Theories abound in which bipedality's upright posture exposes genitals and invites copulation; male ornamentation (as in peacocks' tails) influences females' choice of mates; the costliness of useless ornaments becomes a sign of fitness (i.e., virility to spare); the right shade of red in birds' feathers is a turn-on. Geoffrey Miller's *The Mating Mind* carries this even further, treating the intelligence generated by gradually enlarging brains as a sexual come-on involving the talk, music, painting, ornament that superior brainpower produces as an aphrodisiac—plausible up to a point, but to suppose that Beethoven's last quartets are just a ploy of evolution to attract females is quite a stretch.

What will surprise a newcomer to evolutionary science is the extremely brief period during which a civilization like that of the West has been in existence. Most of the very basic elements of what we now call culture are associated with the Great Leap Forward of forty thousand years ago, the period of the Cro-Magnons, when paintings, carvings, ornamentation "suddenly appear in the archaeological record, together with musical instruments such as bone flutes, and it wasn't long before stunning creations like the Lascaux Cave murals" (35) started off a process that Dawkins sees as a precursor to the Sistine Chapel and the *Goldberg Variations*. Most of the so-called "venerable" traditions that people speak of today— hanging Christmas lights, preserving "family values," idolizing childhood and children, human rights, and standards of "mental health" are merely recent flickers in the evolutionary movie. With writing only five thousand years old and farming only ten thousand, the matters of settling down into communities, growing crops, building houses, establishing legal systems, specializing in artisanal skills—all these are products barely as old as yesterday, a micro-moment in a sequence that will take us back four billion years.

The domestication of animals has changed the genetic makeup of those that were bred for specific purposes, sometimes benignly, as with dogs, which are all derived from the gray wolf no matter how diverse the spectrum from Pekinese to pit bull. Yet settling down has not always produced beneficial alterations in humans, animals, and plants. We learn, for example, that lactose intolerance and allergic reactions to wheat derive from the radically altered diet of post-farming societies, whose ancestors stopped drinking (human) milk at age four and ate few cereal grains until farming's systematic cultivation of the grasses that produce them. And we know from today's obesity crisis that the refined carbohydrates that dominate the manufactured, highly processed Western diet are a recent invention that runs counter to millions of years of primate nourishment, while the "germs" in Jared Diamond's *Guns, Germs, and Steel* are disease sources produced by the increased aggregation of humans into settled societies where transmission of infectious agents becomes all the more likely.

These human interest considerations occupy only about one-sixth of

The Ancestor's Tale, but that is a relatively large portion of the book considering how small a percent of life on earth involved the evolution of hominids. Though the bulk of the book deals with animals, most of these creatures, such as barnacles, worms, and coral, would not occur to laymen as animals at all. Priority in evolutionary sequence, however, is given to funguses, then the rest of the multicellular organisms, then plants, and finally single-celled microbes, the very foundational elements in the origins of life.

One of the most important events in the animal story is the devastation of our planet sixty-five million years ago by a comet that snuffed out not only dinosaurs but about half of all other species. As Dawkins describes it:

> The noise of the impact, thundering round the planet at a thousand kilometers per hour, probably deafened every living creature not burned by the blast, suffocated by the wind-shock, drowned by the 150–metre tsunami that raced around the literally boiling sea, or pulverized by an earthquake a thousand times more violent than the largest ever dealt by the San Andreas fault. And that was just the immediate cataclysm. Then there was the aftermath—the global forest fires, the smoke and dust and ash which blotted out the sun in a two-year nuclear winter that killed off most of the plants and stopped dead the world's food chains. (170)

The elimination of dinosaurs resulted in an amazing proliferation of animal and mammal life forms, formerly nocturnal and very small (to evade the dinosaurs). As I write this in January 2005, newspapers are reporting the discovery of small mammal fossils showing digestive remains of tiny baby dinosaur bones but these seem to be atypical of the ecology of that Cretaceous period, when dinosaurs ruled.

Raising the subject of whether today's technology has the power to intercept similarly life-destroying missiles from outer space, Dawkins cannot resist a few stabs at the Bush regime that he so profoundly loathes: "Politicians who invent external threats from foreign powers, in order to scare up economic or voter support for themselves, might find that a potentially colliding meteor answers their ignoble purpose just as well as an Evil Empire, an Axis of Evil, or the more nebulous abstraction 'Terror,' with the added benefit of encouraging international co-operation rather than divisiveness . . . The mass realisation that humanity as a whole shares common enemies could have incalculable benefits in drawing us together rather than, as at present, apart" (171). This, if anything, could be said to be the submerged thematic undercurrent of an overtly athematic macro/micro zoological history of Planet Earth.

This implicit theme—of the unity of life, so dramatically revealed in the project of tracing us back to bacteria—surfaces again in a major aria later in the "pilgrimage," an account of "ring species" and racism, nicely tailored to illustrate another of Dawkins' favorite ideas, the illusoriness of gaps and discontinuities and the false belief in essences, going back in Western culture to Plato's essentialist ideas of good, beauty, and whatnot. It is worth quoting at length one of the most fascinating passages in the entire book, on ring species:

> If you follow the population of herring gulls westward to North America, then on round the world across Siberia and back to Europe again, you notice a curious fact. The 'herring gulls,' as you move round the pole, gradually become less and less like herring gulls and more and more like lesser black-backed gulls until it turns out that our Western European lesser black-backed gulls actually are the other end of a ring-shaped continuum which started with herring gulls. At every stage around the ring, birds are sufficiently similar to their immediate neighbors in the ring to interbreed with them. Until, that is, the ends of the continuum are reached, and the ring bites itself in the tail. The herring gull and the lesser black-backed gull in Europe never interbreed, although they are linked by a continuous series of interbreeding colleagues all the way round the other side of the world. [i.e., They are separate species.] (303)

If it had happened that humans and chimps were a ring species, "what would it do to our attitudes to other species? Many of our legal and ethical principles depend on the separation between *Homo sapiens* and all other species." As for people who blow up abortion clinics, eat meat, and don't care about chimps, "Would they think again, if we could lay out a living continuum of intermediates between ourselves and chimpanzees?" (303). Much of the illusion of essentially different species is the result of the absence of fossils for the intermediate forms that would show an unbroken link between apparent disparates. "There is no such thing as essence," Dawkins writes. If intermediates were still visible, "instead of discrete names, we would need sliding scales, just as the words hot, warm, cool and cold are better replaced by a sliding scale such as Celsius and Fahrenheit" (308).

The further implications of this continuity are picked up a hundred pages later in a discussion of race in connection with "The Grasshopper's Tale." In a brilliantly conceived use of a photograph showing Condoleezza Rice, Colin Powell, George Bush, and Donald Rumsfeld standing side by side, Dawkins asks whether a Martian seeing them all together would suppose it was a case of three whites and one black. That Powell is regarded as "black" (standing between a very dark Rice and a white Bush) when he more closely resembles Bush and Rumsfeld, raises all sorts of ethical questions about essences, continuity, and illusions. "Why," Dawkins asks, "do people so readily swallow the apparent contradiction . . . between the verbal statement, 'he is black,' and the picture it accompanies?" It is, he replies, an instance of the "tyranny of the discontinuous mind" (401–2). This gradual blending over large periods of time, concealed by missing intermediate fossils, recurs as ground bass throughout the accounts of forty branchings. Whales, it turns out, are close aquatic cousins of hippos, whose characteristic body parts lie within the whale body; and we ourselves are kin to lobefin fish, "who have muscles in the fleshy fins themselves, just as we have biceps and triceps muscles in our upper arms and Popeye muscles in our lower arms" (330). Species, Dawkins believes, are the illusory fixities of mind-created discontinuities. Biologists regard inability to mate as the criterion for recognizing species but the deeper continuities, like the hippo/whale relationship, are masked by this taxonomic privileging. After quoting a really shocking racist passage from H. G. Wells invoking genocide against blacks and Jews, Dawkins asks, "What, I wonder, will our successors of the twenty-second century be quoting, in horror, from us? Something to do with our treatment of other species, perhaps?" (405). Given the already proliferating writings on the subject of speciesism from Peter Singer to J. M. Coetzee to People for the Ethical Treatment of Animals (PETA), it's a pretty plausible guess.

In "The Fruit Fly's Tale" this theme is picked up yet again in an extended account of the Hox gene, which Dawkins describes as "a gene whose mission in life is to know whereabouts in the body it is, and so inform other genes in the same cell" (418) so that legs do not grow out of heads, as they are sometimes known to do when Hox genes malfunction. Since Hox genes are a characteristic of animals (as opposed to plants and protozoa), Dawkins sees them as yet another type of unity: "The Hox story shows that animals are not a highly varied, unconnected miscellany of phyla, each with its own fundamental body plan acquired and maintained in lonely isolation. If you forget morphology and look only at the genes, it emerges that all animals are minor variations on a very particular theme" (424).

As we go back farther and farther to the earliest concestors, we arrive at the first animals that Dawkins can trace, a category of single-celled parasites known for short as DRIPS. These are preceded in time by fungi, then plants, which "sit, indispensably, at the base—the very foundation—of nearly every food chain . . . the first living things any visiting Martian would remark. By far the largest single organisms that ever lived are plants" (506). Lastly, or firstly, are the bacteria, the earliest and most prevalent form of life on this planet. "There is no doubt that the great majority of life's diversity at the fundamental level of chemistry is microbial, and a substantial majority of it is bacterial" (554). But not until the development of the contemporary science of molecular biology was it possible to inspect the real structure of living things. "We didn't even know about bacteria until the nineteenth century," or even whether the specimens seen through powerful microscopes were animals or plants. Amoebas were once thought to be the "grand ancestor of all life—how wrong we were, for an *Amoeba* is scarcely distinguishable from a human when viewed through the 'eyes' of bacteria" (555). Molecular biology has resolved many of these ambiguities by showing the limitation of morphology, the relatively superficial study of apparent forms.

As we arrive at this point close to the beginnings of life, a number of philosophical questions inevitably suggest themselves. If once again a missile from outer space were to destroy most of existing life, would evolution rerun its course more or less as it did this time around? Dawkins thinks that the rerun would be similar to but not identical with what we now have. To support this supposition, he reminds us of the phenomenon of "convergence," the independent development of similar structures in species that have been isolated geographically. The movement toward eyes, for example, seems inevitable, independently achieved by various isolated species, but a new round of creatures who have them would almost certainly not be identical with the creatures of today, since natural selection depends on ecological and genetic variables that have no real chance of occurring exactly as they have done once before. Similar but different looks like a prudent guess.

Equally speculative is the question of how nonlife became life, though Dawkins equates the first instances of heredity, that is, replication, with "life" itself, since these events initiated the chain whose links are all the species that have in fact evolved. But the mysteriousness of "life" seems less mysterious than sheer existence after reading a book like this, further demolishing the notion of life as a "spirit" breathed into matter by some sort of transcendent bellows. More intelligible, more plausible (to me, at any rate) is the recognition that chemical reactions and physical changes, the intermixing and clinging together of elements and their subsequent transformation into greater complexities, are not many steps away from the simplest form of "life." One can easily imagine a single jog that jolts these chemico-physical reactions into a self-sustaining process to be known as "life." Or as Dawkins put it in *The Blind Watchmaker,* "There is nothing special about the substances from which living things are made. Living things are collections of molecules, like everything else."[2] Beyond this, the development of consciousness and self-consciousness strikes me as much more unfathomable than mere "life." The cognitive neurosciences still have a long way to go to psych out psychology itself. Life begins to look more and more like physics and chemistry taken to a point of even *more so.* But consciousness?

Dawkins' final pages, which he calls "The Host's Farewell," are an expression of wonder as to why there is something rather than nothing and why that something is sometimes "life." "The fact that life evolved out of nearly nothing, some 10 billion years after the universe evolved out of literally nothing—is a fact so staggering that I would be mad to attempt words to do it justice" (613). Taking a parting shot at cheap simplistic supernaturalism that explains nothing beyond human fantasies and desires, he concludes, "My objection to supernatural beliefs is precisely that they fail to do justice to the sublime grandeur of the real world. They represent a narrowing-down from reality, an impoverishment of what the real world has to offer" (614). And the real world of *The Ancestor's Tale* is, in a word, fantastic.

PART THREE / Consciousness

CHAPTER TWENTY

Muses, Spooks, Neurons, and the Rhetoric of "Freedom"

For my part, when I enter most intimately into what I call *myself,* I always stumble on some particular perception or other, of heat or cold, light or shade, love or hatred, pain or pleasure. I never can catch *myself* at any time without a perception, and never can observe anything but the perception . . . If anyone, upon serious and unprejudiced reflection, thinks he has a different notion of *himself,* I must confess I can reason no longer with him . . . He may, perhaps, perceive something simple and continued, which he calls *himself*; though I am certain there is no such principle in me.

DAVID HUME, *Treatise on Human Nature* (1739)

Muses, Spooks, Neurons

In the beginning, the ancients talked about the Muses; later on, Milton spoke of the Creator Spiritus; while Yeats had us rolling on the floor when he spoofed us with spooks, who brought him images for his poetry through what he called, more aptly than he realized, automatic writing. But all of them were onto something: they realized they hadn't a clue as to where their creativity came from; it all seemed so magical, so implausible, so involuntary. For how could a self freely will ideas, metaphors, images or anything else into existence? To will them, it would have to know them, to hold them all in an omnipresent memory—or at the very least hold them in a mega-index, a veritable Google, of all the brain's contents, which it would need to know by heart—and experts report there are at least 50 billion neurons, and maybe 150 billion—storing the data of our lives. But "we" don't choose the items in our so-called stream of consciousness anyhow—they come unbidden, they "enter our mind," so what good is an index?

As for the self, forget about anything more than a virtual self or a self effect, unless you can entertain the idea of a spooky homunculus dwelling in the pineal gland, a "central meaner," to use Daniel Dennett's derisive term, who watches the movies, the stream of consciousness, being shown in the brain's Cartesian Theater, and pulls it all together into meaningfulness.[1] An organizing center of consciousness, moreover, would make the rest of the brain superfluous, since the center, the "I," would already be a brain unto itself, knowing everything we attribute to the myriad other faculties, for how else could it request their data? So a self would entail an infinite regress of explanations to account for *its* knowledge, a pre-self that tells the self what it knows, and a pre-pre-self to inform the pre-self, and so on. And the free will that supposedly animates the imaginary self? It can only be an oxymoron. How could a purely virtual self, empty of motivations, no more substantial than the projected image of a movie onto a screen, make choices from the blankness needed to be "free" of predispositions? "Choices" stem from *unchosen* motivations, and the "self" may already have billions of them. (Do "we" choose who will arouse us sexually, what foods will stimulate our appetites, what thoughts will enter our heads?) Without motivations, there wouldn't be any behavior at all. Why get up and cross the room for no reason at all? Why express a thought that just presented itself? Why go into a rage when nothing has happened to send you into one? Daniel M. Wegner in *The Illusion of Conscious Will* has a great deal to say on this subject. [2]

> The fact is, each of us acts in response to an unwieldy assortment of mental events, only a few of which may be easily brought to mind and understood as conscious intentions that cause our action. (145)

> We perceive minds by using the idea of an agent to guide our perception. In the case of human agency, we typically do this by assuming that there is an agent that pursues goals and that the agent is conscious of the goals and will find it useful to achieve them. All this is a fabrication, of course, a way of making sense of behavior. (146)

> Actions and their meanings are stored separately in memory. Otherwise, we would always know exactly what we intend and never suffer the embarrassment of walking into a room and wondering what it was we wanted there. (166)

[I]ntentions are often matters of self-perception following action, not of self-knowledge prior to action. (175)

This position becomes all the more cogent in view of empirical findings of the cognitive neurosciences to the effect that desires are registered in the brain microseconds before they occur to consciousness and will, a phenomenon of which Wegner, among others, gives an account. Since the publication of Wegner's book in 2002, the matter has been presented in minute physical detail by David M. Eagleman in an article appearing in 2004: "Human brain studies using electroencephalography (EEG) have long suggested that some part of your brain was already moving toward [your] decision well before you were aware of it," that is, before you were aware of making the decision![3]

As for "free will," it could only mean being constrained into stasis by sheer vacuity, the freedom to be an undifferentiated lump with no predispositions, certainly no unconscious "intentions" that one simply acts out and later explains. Free will, an absurd concept that has been bandied about for centuries, would be no will at all, unconscious or otherwise.

Just what, you may ask, *are* those neurons, aka "brain cells," to which everything is attributed these days and whose connection with the mind John Donne, to the best of his knowledge, tried to account for hundreds of years ago as "that subtle knot that makes us man," the nexus of body-mind interaction?[4] The conjunction of neo-Darwinism, evolutionary psychology and biology, the cognitive neurosciences, and the extraordinarily sophisticated machines for brain exploration devised by high technology has thoroughly and finally overturned—except in popular culture and religious superstition—the Cartesian dualism that dominated Western thought for so many centuries. Those billions of miniscule brain cells interact by means of their threadlike dendrites and axons to produce one million billion connections known as synapses. Neurons are said to "fire"— by which is meant that an electrochemical transmission takes place across these synapses when triggered by stimulation from anywhere in the body and from different areas of the brain itself.

One of the great mysteries thus far unsolved is the nature of neuronal storage. Although the neurons are referred to as "the message-carrying cells of the brain," just what is it that's being carried or stored? The consensus is: neither images nor "ideas," which is to say that the brain is not a repository of representational materials. Unlike a vinyl LP inscribed with a visual analog of the sound that has been recorded—a real "representation"—the neuronal storage of the brain is like a digital compact disk, insofar as the storage medium (ones and zeros in the case of CDs) has no resemblance to what is finally produced (a "scene" in consciousness), only coded instructions for producing it. This effectively destroys the classical dualism of a head full of pictures, ideas, and sense data that are organized and overseen by a little guy inside. (And how was *his* consciousness presumed to derive *its* content? From littler guy number two, ad infinitum?)

The model now holding sway in the neurosciences establishes that consciousness is produced ad hoc over and over again each microsecond by some of the million billion synapses that undergo incessant firing. The smooth continuity of conscious experience is an illusion like one's sense of the continuity of sound produced by forty-four thousand digital samplings per second in the recording and playback of a compact disk. Just as our brains are not constituted to detect the forty-four thousand interruptions of sound between the samples, it is not constituted to be aware of the pointillistic or pixilated (I grasp for words here) nature of consciousness, any more than it can notice the saccades—constant jerky refocusings of the eyes in vision—that finally yield an apparently smooth visual panorama. Even in the case of memory, we do not re-view old images that are merely retrieved from the brain as whole cloth in acts of recollection. Rather, what is recalled is reconstructed ad hoc from its electromechanical source, producing results that are never exactly the same twice. How this constant movement, firing, and assembly produces final coherent awareness remains as mysterious to us today as the workings of John Donne's subtile knot.

As Gerald Edelman phrases the question: "How can it be that, despite the absence of a computer program, executive function, or superordinate map, up to thirty-three functionally segregated and widely distributed visual maps in the brain can nevertheless yield perception that coherently binds edges, orientations, colors, and movement into one perceptual image?"[5] If he is unable to supply a definitive reply, one thing seems pretty clear nonetheless, to use Edelman's words: "The world is causally closed—no spooks or spirits are present—and occurrences in the world [i.e., the activity of the body] can only respond to the neural events constituting C' [the neural actions of the brain]" (78–79). Or as Daniel Dennett succinctly puts it: "Where does it all come together? The answer is: Nowhere . . . There is no one place in the brain through which all these causal trains must pass in order to deposit their content 'in consciousness.'"[6]

This picture of consciousness is hardly revolutionary. One reads with astonishment the prescient account given by Thomas Henry Huxley with the vocabulary available to him in 1874:

> It is quite true that, to the best of my judgment, the argumentation which applies to brutes holds equally good of men; and, therefore, that all states of consciousness in us, as in them, are immediately caused by molecular changes of the brain-substance. It seems to me that in men, as in brutes, there is no proof that any state of consciousness is the cause of change in the motion of the matter of the organism. If these positions are well based, it follows that our mental conditions are simply the symbols in consciousness of the changes which take place automatically in the organism; and that, to take an extreme illustration, the feeling we call volition is not the cause of a voluntary act, but the symbol of that state of the brain which is the immediate cause of that act. We are conscious automata, endowed with free will in the only intelligible sense of that much-abused term—inasmuch as in many respects we are able to do as we like—but none the less parts of the great series of causes and effects which, in unbroken continuity, composes that which is, and has been, and shall be—the sum of existence.[7]

Although Gerald Edelman cites Huxley with approval, he does not go along with the automata remark. Rather, he concludes: "The very richness of core states provides the grounds for new matches to the vicissitudes of the environment" (85), rejecting the view that people are simply "Turing machines," that is to say, computers.

Against a background such as this, it is instructive to reconsider one of the shopworn truisms of the past fifty years of orthodoxy in the humanities. Roland Barthes and Michel Foucault, for example, wrote about the disappearance of the author, but all they really meant, in keeping with their Marxian line, was not that the writer is dissolved into these neurons but that the writer, like everyone else, is socially constructed, simply a mouthpiece for social programs: in brief, the Standard Social Science Model, innocent of the evolved brain-centered predispositions from millions of years of primate and hominid existence, as described by Leda

Cosmides and John Tooby in their landmark article, "The Psychological Foundations of Culture."[8] But the cognitive neurosciences suggest something far more radical: the author as a "self," like every other kind of self, really doesn't exist at all. The author is not *constructed* by society; rather, "the author" is constructed by billions of involuntary neurons with a vast prehistory, constantly *reformulated* by culture. The author, as a disembodied self, a locus of creativity, is a phantom. Maybe we should speak of "the author effect" or the virtual author.

To flesh this out as an author—or author-effect—I would like to do a sort of Cartesian meditation based on two highly creative critical essays written by the virtual "me" as virtual author. Of the many retrospectives I have written on a writer's oeuvre, the one called "Sylvia Plath, Hunger Artist,"[9] like many of the others, was set off by the publication of a new biography, although three or four already had been published. Typically, to produce a retrospective essay, "I" go through three stages: reading/research, thinking, writing.

So for starters, I had to read the three previous biographies before the newest one, then all of Plath that I had not previously read (including her journals and letters home to her mother), and finally a generous sampling of the critical literature. Fine. But who was the reader of all this? An imperial, self-determining but somehow unpredisposed blank consciousness that applied its empty, undesiring self to totally transparent texts that simply said what they mean? Transparent to whom? The virtual self doing the reading can only read through an immensity of accumulated baggage, sometimes referred to as "filters." This baggage consists not only of genes but of the year and place of one's birth, one's parents, one's education (all the books one has ever read, all the movies seen, music heard, newspapers and websites examined), the time and place in which one lives, one's temperament, bodily states and infirmities, desires, tastes, traumas, joys. Had I been born only twenty years ago, I would doubtless have a very different response to nose rings, hip-hop, TV, computer games, Beethoven's quartets, Albert Camus, Lionel Trilling, and so forth, and the lenses through which I read the Plath writings would have revealed very different texts from the ones I actually experienced as a venerable me. This is not to take the deconstructive view that there is no text but only to suggest that texts are instructions for reading, not finished substances, and your ability to follow the instructions depends on the contents and disposition of your mental storage facility. Had I been born in Iraq and learned English there, would my intellectual, emotional, esthetic, and moral responses to Anglophone literature be the same as they are now? In sum, unbeknownst to me, my takes on everything I read for the Plath essay had already been inscribed in my neurons for potential involuntary retrieval later on as thinking and writing.

As for the thinking, a lot of it is going on as my virtual self reads all the books through its accumulated filters or baggage. What I take in, what I dismiss, what touches deepest areas of my psyche—to some degree I am aware of it, or there is an awareness of it, if not from a me. But when the reading is finished, as I mow the lawn, drive the car, or brush my teeth while consciously ruminating about Plath's life and works, a whole other order of thinking is going on in another inaccessible realm about which I haven't a clue. This order will not come into obvious play until I sit down at the computer and watch, with a certain sense of surprise, the seemingly unpremeditated thoughts that pour out onto the page. Or, to borrow a question from E. M. Forster's *Aspects of the Novel,* "How do I know what I think until I see what I say?" And if the "I" is not the cause but the effect of inscripted neurons, it can be said without too much self-disparagement that as a writer, "I" am truly asleep at the wheel, a dreamer, the driven rather than the driver, taken for a ride by an autopilot.

As for the writing process, there are two types of writer I am aware of: Type A sits in an easy chair with a yellow legal pad and pencil, eking out, sentence by agonized sentence, over the course of one disciplined day after another, a page or two of writerly produce at each long session. Type B, and that's me, couldn't possibly produce more than a few uninspired sentences by sitting down at the computer in cold blood. Like the high dew point that triggers monsoons in Tucson, my home, a moment of high psychological "do" point is required for me to sit down at the keyboard and precipitate. But when that moment arrives, a torrent of prose starts to flow from my head, through my fingers, onto the keyboard, amazingly coherent and organized though nonetheless needful of editings to produce a final draft. After about an hour or two of this, I find myself completely exhausted, drained, unable to write another word. A good day is two pages. The next day, I pick up where I left off, first doing some editing of the previous day's output and then continuing on for another couple of hours until debility sets in. Eventually, I have an almost final draft, a complex web, full of interrelations, allusions, ideas I didn't even know I was thinking, a point of view that sometimes takes me by surprise. So *this* is what I thought of the writings of Sylvia Plath! So *this* is what I made of those four biographies! And the overriding question is pretty clear: where the hell has it all come from? Not from "me," for a phantasm can't generate a thing. How could "I" have been the creative intelligence that produced the final essay when "I" had only the vaguest premonition of what those unmannerly, secretive, and unforeseeable neurons were up to? What they don't "tell" me I can't know.

And where did the actual sentences come from? Not from consciously measured prose poured from precisely calibrated alembics under "my" control. But from Chaucer, Shakespeare, Herrick, Thomas Browne, Fielding, Pope, Sam Johnson, Austen, Arnold, Virginia Woolf, Tom Wolfe, Bellow, and all the rest of them, not to mention academic criticism and philosophy, scholarly journals, newspapers and periodicals, pop culture, a whole range of models that infiltrated my psyche unbeknownst to "me." A lifetime of unwitting absorption of syntaxes and dictions, locutions and lexicons (can you hear that convoluted echo of Sir Thomas Browne, the Miltonic periodicities of Wordsworth, the obstreperancies of Bernard Shaw, the demotic epithets of Tom Wolfe?), a workshop of molds into which my thoughts effortlessly conformed themselves. Born twenty years ago and brought up on TV, the Web, and the movies, not even reading the newspapers, would I sound more like a dithering Ozzy Osbourne and his wacko family? Or am I an imperial self-directed consciousness, an image of God, freely willing "somethings" out of "the nothing"?

In 1996, after a three-thousand-mile auto trip from my home at that time in Chicago's suburbs to Tucson to L.A., Sacramento, Reno, and back to Chicago, I experienced an uncanny case of what used to be called inspiration from the Muses that issued in the essay "Ecology and Ecstasy on Interstate 80" (chapter 10 of this book).[10] A contrapuntal interweaving of ecology, technology, and the arts, it drew to a close as I described my homebound traversal of Iowa on I-80, listening to the Metropolitan Opera, observing the budding landscape of early spring, and thinking about the technology that was bringing *Die Walküre* to me live from New York in my speeding car. One particular sentence, actually a mere phrase, from the final page can serve as a startling instantiation of my thesis, namely (in case you missed it) that the creative writer (or creative anybody—or just any old slob) is a passive agent of neuronal storage, energy, and organization. It reads as follows: "As I reached the Quad Cities area and began the crossing of the Mighty Mississippi—with due regard for the river-defining technologies of Mark Twain, T. S. Eliot, and the Army Corps of Engineers—signs of spring were definitely in evidence, buds were opening, the air was warming, and Sieglinde was singing the most rapturous passage in all of *The Ring.*" The river-defining technologies of Mark Twain, T. S. Eliot, and the Army Corps of Engineers? Fantastic! Who could have thought up such a thing? It was conceived and written in a matter of seconds as part of my daily hour or two of manic output and was definitely not teased with effort out of my billions of neurons by a brilliantly Googliferous "me." Just where on earth did it all come from?

About forty years prior to the act of composition, I had read some of Mark Twain's writings about his early life as a pilot on the Mississippi and the technological skills that were needed to do the job. I had neither read nor thought about these writings ever since. Between twenty and thirty years before 1996, I had read Eliot's *Four Quartets* several times, always associating his words about the river as a great brown god with his birth in St. Louis and his early proximity to the Mississippi, which is not mentioned by name. And perhaps half a dozen years before the day of composition, I had read John McPhee's brilliant book with the ambiguous title *The Control of Nature,* one of whose three essays was about the ongoing but ultimately futile attempt of the Army Corps of Engineers to keep the Mississippi from wandering out of its government-approved riverbed. No consciously controlling "me," however brilliant, could possibly have ferreted out these deeply buried and scattered inscriptions in my brain, old thoughts that had not surfaced in years but that now organized themselves within seconds to serve the turn of the automatic writer who was producing these sentences. Those dusty old neurons are smarter than "I" am, putting Google to shame.

Am I going so far as to say that what I write is totally outside of "my" control? Not exactly: "I" see what "I" have written, it gives "me" new ideas, improvements are possible, changes can be made (these quotes around pronouns are becoming oppressive). But when I read my own writing, when I see what I have written, I am still the reader reading through his own filters, carrying his own baggage. The thoughts for improvement that occur to me are as much the product of my neuronal history as any other thoughts. That's why I need feedback from my friends, who read through different filters. Do I feel trapped within involuntary paradigms? Certainly not. Without a psychological history and an affective shape I'd be nothing, a vacuous airhead. The desire to improve my performance is not a self-created desire, since the virtual self can't create anything; as an insubstantial phantasm it has no powers of agency at all. Should I be taking credit, then, for writings "I" didn't write? Strictly speaking, no. But everyone else is doing it. Why not me? We are all playing the same game because there's no other game in town.

The Rhetoric of "Freedom"

I will be asked if I think people are trapped in changeless paradigms not of their own choosing. Not of their own choosing, yes and no: it depends on what you mean by "their." The self cannot initiate a thing, because it's an effect, initiated elsewhere like characters on a movie screen. On the other hand, we're changing all the time, growing, learning, maybe improving. But the will to change is not generated by a self. It comes from involuntary sources, either internal or external, though even external sources are finally internal, inscrutable, neuronal. When I eat a tasty new dish at a restaurant, there is no "I" that determines (in a causal sense) that the food is pleasing to me. "I" find, discover, recognize it, as pleasing. There is no choice or willing involved. Nor when I respond sexually to person A but not to person B is it a matter of choice: I recognize the tingles in the flesh, the heat, the increased heart rate, the stirrings in the genitals, and my behavior is influenced accordingly toward A but not toward B. When a friend gives me a lecture on how I should respond to a poem, a person, a theory, there is no "I" that wills the lecture to be convincing. Rather, I find myself convinced or unconvinced. If a certain odor of flowers, of sweat, or of food should happen to accompany one of these lectures by sheer fortuity, a reinforcement or repulsion totally unwitting to me could very well take place that "I" have nothing to do with.

What, dear reader, is convincing you that I am mounting a persuasive argument right now—or that I am merely full of crap? Is it because you are a superior intelligence, a scholarly reservoir of judicious knowledge, an infallible sensibility that is *choosing* to find me convincing or not—or because your private biochemical planets happen to be in the right conjunction for receiving or rejecting this particular attempt of persuasion? After all, yesterday's purportedly loony tunes are now fully naturalized, and they "go without saying," even for you, who no longer make disparaging remarks about blacks, Jews, women scholars, or queers. This change in your attitudes is not the result of your superior moral intelligence or sensitivity, of your splendid transcendent consciousness making enlightened choices. It's the cultural ethos, the ambient air, the mental environment of your particular cohort as it all does its work along your domesticated wires—and now you "find yourself" with newfound enlightenment. (Bully for "you.") Teenagers born after your dispensation are "enlightened" in many of these matters from the start. Is that because they are wiser and more judicious than you? Capable of making more civilized "choices"?

I want to call these generative forces—and many others—rhetoric, going far beyond classical ideas about persuasive speech. There is a world of unwitting persuasions, not in the sense of carefully articulated vehicles of writing or speech (which produce their own powerful influence) but a word here, a gesture there, a new friend, a phrase in the newspaper, a traumatic event, all totally serendipitous, that alter our baggage, reconstitute the screening mesh in our filters, changing our perspectives and desires. These are relatively "conscious" rhetorical messages, but persuasions extend far beyond these to realms exceeding our grasp, realms that do not involve a *composer* or a sender but only a *receiver* of persuasions via an indigenous system of neuronal rhetoric (a replacement for the synthesizing little Cartesian guy in one's head) whose logic we are only dimly able to discern. It is the machinery of reception that produces this rhetoric, turning unwitting and chaotic experience into meaningful messages or narratives. (Think, for example, of the elaborate narrative structures of dreams. Is the waking "stream of consciousness" any more under the control of an agent-self than those?)

Contrary to the postmodern assumption that there is a knowing subject whose total experience consists of acts of textual (in the broadest sense) understanding, the knowing subject is as much a fiction as the self, a putative free-floating consciousness that is somehow disconnected from an elaborate neuronal system to which everything else is addressed, not in the form of constructed rhetorical "texts" but as smell, touch, sight, sound, tastes, events, as well as words, not to mention internally generated feelings, with a life of their own apart from conscious thought and from which consciousness emanates. The real addressee of this rhetoric is not a deciding, weighing, subject/self/person that is "persuaded" by textualized ideas (whether verbal or otherwise) but an involuntary, inscrutable system of preconsciousness that generates meaning. (Freud had a go at it but his mythologies required more faith than the Holy Ghost.) This system is now slowly being made visible as brain activity by means of advanced technological procedures such as MRIs, corroborating Daniel Wegner's claims about the disconnect (and delay) between the initiating brain activity and the illusion of willing attributed to virtual "selves" and "subjects" who attempt to produce ex post facto explanations of what they have experienced. Phenomenal experiences are processed as a neuronal rhetoric, laden with meanings, ideas, and affect, as persuasive as carefully composed speech or writing, if not more so.

The events of 9/11, for example, have had the profoundest effects on the feelings, attitudes, deepest emotional core of millions of people throughout the world, unalterably changing their psyches. The destruction of the twin towers was, in my sense, a rhetorical event, triggering a vast spectrum of altered "being in the world" (in the Heideggerian mode). The panoply of television programs, sitcoms, video games, sports, commercials, images, events, sounds has the profoundest of effects on the sensibilities of children, pubescents, and adults who view them. The sordidness of corporate CEOs, the mendacities of the Bush regime, the photo of a corpse on a dusty roadway in Iraq, a sentence that slipped out of the mouth of a friend that turned one against him forever—the examples are infinite. These rhetorical triggers generate the psychological evolution of human societies and individual persons. (Richard Dawkins calls them memes but his working out of the concept strikes me as an imperfect analogy with genes.)

I want to connect these triggers with the idea of human "freedom," while retaining the point of view that "free" is a nonsense term suggesting that something can exist outside a chain of causes or take place without antecedents. "Freedom" in that sense would be madness, chaos, moral entropy. We say we "know" other people—because they are more or less predictable. To have a character, to be relatively predictable (at least enough to qualify as a "person" or a "thing") is the nature of sensible materials that exist in time. If I *know* that Phyllis will like this dress I am buying for her, I also "know" that Phyllis is not "free" to like just anything, that Phyllis is a congeries of qualities that excludes other qualities. She is a determinate selection from the totality of material possibilities. If she were "free" she wouldn't be Phyllis at all but, rather, an amorphous lump of nothingness—and I wouldn't have a clue about what to buy her. (Of course, I can never know her totally, since there is no totality in an open-ended process.)

We complain when people are unfathomably unpredictable, we whine that they lack "character," that is, determinateness, since to be, *for us,* is always to be *as something*—and something that we can recognize, categorize, domesticate. *But human unfreedom is the very condition of personhood and identity.* Rhetoric, here extended to include the unwitting reception of happenstance stimuli that generate formalized neuronal narratives, goes to work on all this determinate material, altering, developing, and enriching it. This enrichment, this complexification, is as close as I am able to get to the idea of "freedom." Through the power of a rhetoric unbeknownst even to ourselves, we are able to influence and change the behavior of other people (and be changed in turn by them) even if "we" are unable to change our own behavior. Thus human beings are "free" to become almost anything their evolutionary inheritance permits—but not anything "they" wish, since the changes are not initiated by a self. *Homo sapiens* have thus far evolved from slime to primates and thence to complex, urbane (and sometimes murderous) bourgeois. The possibilities look open-ended, but "you" can't will them into being for yourself. You don't even know what they are or could be. People are changed, unwittingly, willy-nilly, not by "self," not by choices, but by rhetoric, in my expanded sense, in the past delivered by muses or spooks but now understood as narrative-generating neurons, as rhetorical as anything produced by Ovid. Think of Richard Dawkins' cosmic blind watchmaker now incarnated phenotypically as the dumb rhetorician.

What other sense of freedom would you like: unpredictable, unmoored, madcap? People are, after all, unpredictable enough, despite all their character-defining predictability, simply because we can never know the ultimate springs of their behavior, what those neurons are up to. Why am I so willing to sit here at my computer and tap this out, what inscrutable need or drive is goading me on? No shrink on earth will ever be able to solve it: his explanation will be the "rhetoric" of a fashionable paradigm and its narrative will either strike me as convincing or strike me as specious, not because he is "right" or "wrong" but because my unfathomable biochemistry is imprinted in a certain preconscious way that accepts or rejects his story while producing the ongoing "I" that is "me," which itself has a zero degree of agency.

In sum, just as my literary and critical creations are the music of my neurons (formerly the Muses), so is human behavior in general a form of rhetorical music, sometimes heard, more often unheard—but heeded nonetheless. Our behavior is altered by what we hear and read, by what we see, feel, think, smell, however involuntary the alterations may be. We are surely not doomed to a simple repetition of the past, however much our primordial constitutions may lay the tracks on which we move forward. Are we "trapped" and "unfree"? What could *untrapped* and *free* possibly mean? To be is already to be "trapped" in a particular configuration of matter. Are we robots, then? you ask. What would you have me say? Can you even conceive of an alternative? If we're robots we're pretty damned brilliant robots. What more could anyone possibly want?[11]

John Searle and His Ghosts

WITH THE OBSOLESCENCE of traditional metaphysics and its attendant epistemology or "theory of knowledge," the "transcendental" reason that once enabled philosophers to solve cosmic problems from their armchairs without even looking out the window is now being supplanted by laboratories of cognitive scientists. And the information derived from them is becoming foundational for many other disciplines as well. One of the seemingly eternal dilemmas waiting for solution on the conveyor belt of this new epistemology is the problem of consciousness. What on earth is a thought and how does a body produce it?

This crux—what is consciousness?—is "worried" (like a ball of yarn by a cat) by a whole range of contemporary disciplines. The cognitive neurosciences, represented by thinkers such as Gerald Edelman and Walter Freeman, give us highly technical maps of the brain showing neurons, electrochemical forces, inputs and outputs, recursive transmission paths, and so forth, but are always hitting, so to speak, the brick wall of consciousness. Professors of psychology, such as William Calvin, Steven Pinker, and David Barash, produce eminently readable books for more general audiences, fusing the disciplines of traditional psychology, physiology, and biology, with breakthroughs from the neurosciences and insights from the humanities. At the edge of this group is Daniel Dennett, a professor of philosophy with strong enough ties to the cognitive neurosciences to produce a book titled *Consciousness Explained*[1] that candidly confesses its inability *really* to explain consciousness while it sheds all sorts of light nonetheless. Now John Searle, an analytic philosopher with roots in speech act theory leading to a series of books about "mind," has produced *Mind: A Brief Introduction*.[2] Although I have seen a description of this book as suitable reading on a transatlantic flight, I would suggest that a round trip from Chicago to New Zealand might be more appropriate. The book is relatively "brief" but nevertheless requires the kind of serious attentiveness that takes real time.

Searle, who writes a strikingly lucid prose of chatty informality, provides a rapid overview of the ways in which philosophers, principally Descartes, tried and failed to bridge the mind/body divide. From there he goes on to deal with the three main problems of contemporary philosophy of mind: consciousness, free will, and the self. As a "realist" philosopher who believes there is not only a real material world but that human beings are capable of saying true things about it, he is dedicated to providing spook-free explanations of these major problems of the cognitive sciences and philosophy. But as an analytic, "linguistic" philosopher, he depends very heavily on redefinitions of phenomena that tend to eliminate the problems he has set out to solve. As things turn out, carving "reality" into highly finessed, albeit demotic, sentences can carry us only so far. And spooks kicked out the front door have a nasty way of sneaking in again from the rear.

Since the aim of his book is to give a final blow to the age-old positions of dualism and materialism that he regards as obsolete, Searle wants to account for consciousness as produced from the same materials as everything else— i.e., physical microparticles of various types—while recognizing its unique character. Dualism posits immaterial spirits, souls, selves, and thoughts with the preposterous ability to initiate bodily actions and survive physical death (how can a thought move my arm—and what's a "thought" anyhow?), while materialism reduces the mind to a computer executing built-in programs, eliminating consciousness altogether. Years earlier, Searle administered a devastating blow to fanciers of artificial intelligence who attempted to account for human consciousness in terms of computers. In his famous "Chinese Room" fable he imagined himself as a human being who could correctly answer in Chinese questions posed in that language even though he didn't understand a word. He could perform this feat by using a rule book that showed him how to manipulate Chinese symbols without knowing what they meant. Contrasting this performance with his ability to answer questions in English, Searle remarks, "In English, I understand what the words mean, in Chinese I understand nothing. In Chinese, I am just a computer . . . The computer operates by manipulating symbols . . . whereas the human mind has more than just uninterpreted symbols, it attaches meaning to the symbols"

(90–91). Understanding is a function of consciousness, which is not simply a material program. Thus, if neither immortal souls nor a purely algorithmic program are satisfactory explanations of consciousness (i.e., neither dualism nor materialism), what finally can Searle offer as an explanation?

"All forms of consciousness are caused by the behavior of neurons and are realized in the brain system, which is itself composed of neurons," he writes, *but* "conscious states, with their subjective, first-person ontology, are real phenomena in the real world. We cannot do an eliminative reduction of consciousness, showing that it is just an illusion. Nor can we reduce consciousness to its neurobiological basis, because such a third-person reduction would leave out the first-person ontology of consciousness." Despite this inhospitality to reduction, conscious states "have absolutely no life of their own, independent of neurobiology" (112–13). So how are we to reconcile this apparent paradox or contradiction?

Writing in italics, Searle concludes, *"There is no reason why a physical system such as a human or animal organism should not have states that are qualitative, subjective, and intentional* [i.e., focused on something]" (118). "There are not two different metaphysical realms in your skull, one 'physical' and one 'mental.' Rather, there are just processes going on in your brain and some of them are conscious experiences" (128). And for good measure, he adds, "There is no problem in recognizing that the mental qua mental is physical qua physical. You have to revise the traditional Cartesian definitions of both 'mental' and 'physical,' but those definitions were inadequate to the facts in any case" (118). And the solution?

Not very convincing, I fear, at least to an upstart outsider like me. After all of this strenuous philosophy-speak, what have we really learned? Searle calls his solution " 'biological naturalism' . . . because it provides a naturalistic solution to the traditional 'mind-body problem,' one that emphasizes the biological character of mental states" (113). Just as H_2O molecules produce in aggregate a second order substance of "water" whose liquidity seems to have no obvious relation to its substrate, so, Searle claims, consciousness is a subjective, first-person, qualitative second order of neuronal activity, produced by but different from its generative materials. Searle keeps reminding us that absolutely everything else in the world is third person, that is, at least in theory potentially witnessable. Subjective consciousness, on the other hand, is the unique exception because of its inobservability by anybody except the subjective first person himself. But Searle never substantively deals with what "first person" actually consists of in a third-person world, the very crux he's been trying to demystify. The best he can do is to say that first person, or subjectivity, is "just *the state that the brain is in*" (208). And if it is just a brain state (whatever that means), then we have been given little more than a tautology: consciousness is first person is subjectivity is consciousness. "First person" and "subjectivity," it appears, are little more than new names for the infamous old ghost in the machine. Can it be exorcised by anything as simple as a renaming?

Searle then turns to the problem of "free will," reviewing some of the possible senses of that term. Human beings are "free" to make choices, in a limited sense, if they are not being constrained by outside forces (i.e., not bound and gagged). But how constrained are choices when considered from the inside? Early on, Searle states "that all of our psychological states without exception at any given instant are entirely determined by the state of the brain at that instant," yet he is willing to introduce a somewhat mushy term he calls "psychological freedom." That is, while admitting the ways in which rage, hunger, etc. influence our psychological states, he finds them "not in every case causally sufficient to determine the subsequent action." This is so vague and "metaphysical" as to appear pointless, because Searle next admits "that the neurobiology [the action of neurons] is at any instant sufficient to fix the total state of psychology [i.e., one's mental state] at that instant" (227). Since Searle's theme throughout this book is that causation is a closed material system with no intervention possible by spooks, since he repudiates dualism's belief in mental substances (i.e., souls), his conclusions (you would suppose) have pretty well been set by everything that comes before. But then, straight out of left field, he asks, if "free rational decision making is an illusion," why has evolution endowed us with this totally irrational belief that we have somehow escaped the forces that drive everything else? "We really do not know how free will exists in the brain, if it exists at all . . . But we also know that the conviction of our own freedom is inescapable." In sum, "there is still the question of whether or not we really do have freedom" (234–35).

It's pretty astonishing that with all his analytical powers Searle should end up sounding like a religious conservative evoking intelligent design.

Evolution has indeed endowed us with all sorts of characteristics that are counterproductive of both life and truth in the interests of reproduction. Sexual selection (see, among other books, Geoffrey Miller's *The Mating Mind: How Sexual Choice Shaped the Evolution of Human Nature*)[3] has evolved peacocks with immense useless tails, animals with heavy and debilitating antlers, birds with conspicuous colors, and other features and behaviors that attract females—but also attract predators. Many of these characteristics, mainly in males, are now regarded as fitness indicators that lure females to mate, "fitness" meaning reproductive potential and not survivability beyond fecund copulation (leaving aside creatures that evolved to die upon mating). David Sloan Wilson's *Darwin's Cathedral*[4] develops the view that religious beliefs, regardless of their irrationality and falseness, have survival value for groups and cultures. Evolution is not interested in "truth" but in perpetuation of genes.

With all of the analytical clarity that pervades his book, Searle fails nonetheless to examine the very idea of free will. If the concept—and what can it mean other than unmotivated behavior?—is nonsensical and self-contradictory, to ask whether we "have it" is to ask a preposterous question. (Is the moon's green cheese edible? Only yes or no answers will be permitted.) The pointed question is not whether we "have it" but whether it is a concept that makes any sense at all. (Would unmotivated behavior be a sign of "freedom" or dementia?) Perhaps he would have fared better had he not saved his discussion of the self until the final chapter.

"So what then is this self? I think Hume is absolutely right: there is no experience of this entity, but that does not mean that we do not have to postulate some such entity or formal principle" (295). As things turn out, however, the self is not really an "entity," for as an entity it would have to be a spook, an immaterial ghost. The self, for Searle, turns out to be pretty much the same as our everyday banal conception of the self, a sense that there is a continuing "I" behind our consciousness. (Do we need a philosopher to tell us this?) Like the sense of free will, this sense of an "I" goes against the nature of every other substance in the known universe. There are no free-floating nonmaterial "entities" drifting about the cosmos. Searle's conclusion is, "There is a formal or logical requirement that we *postulate* [my emphasis] a self as something in addition to our experiences in order that we can make sense of the character of our experiences" (298). Postulating is one thing. Existing is another. Following this,

Searle tells us that he is not really satisfied with these conclusions. He very well ought not to be.

As an analytic philosopher, Searle has been insufficiently analytical in his dealings with several major problems of consciousness. Had he admitted from the start that the self was purely virtual, a "self-effect," his discussion of free will would have been more pointed and more candid: without a self, the "problem" of free will disappears. If the self is an illusion produced by the body, it has no agency at all. Nothing can come from nothing.

Though Searle's *Mind* is a worthwhile survey of historical positions that have been held in the West about consciousness, and though Searle is very much aware of developments in the cognitive neurosciences, more straightforward accounts of consciousness than Searle's ghost-haunted equivocations are available elsewhere, as we will see in the following chapter on Daniel Dennett.

Daniel Dennett and the Brick Wall
of Consciousness

"Like" and "like" and "like"—but what is the thing that lies beneath
the semblance of the thing?

VIRGINIA WOOLF, The Waves

How could a physical system give rise to conscious experience?

DAVID CHALMERS, The Conscious Mind

Only a theory that explained conscious events in terms of unconscious events could explain consciousness at all.

DANIEL DENNETT, Consciousness Explained

SWEET DREAMS[1] IS BY NO means the book you would want to start out
with if you had never read anything by Daniel Dennett. There are two
distinguished classics in his oeuvre to be read first, *Consciousness Explained* (1991) and *Darwin's Dangerous Idea* (1995), in that order. Dealing as they do with two of the most pressing themes in current philosophy (not to mention certain of the sciences), these books would rank
pretty close to the top of my list of what every twenty-first-century intellectual should know. *Sweet Dreams,* on the other hand, is a slight book
that has been patched together from various talks, articles in professional
journals, and newly written passages, all of which serve to tweak Dennett's major doctrines in the light of subsequent criticisms and rethink-
ings. Unlikely as it may seem, the book reads well—like everything else
by Dennett. It's sheer pleasure to be in the company of a consciousness
like this—if you could believe in consciousness at all after reading what
he has to say.

Still, the basics are hardly in dispute in the matters of self, consciousness, and free will, given the extraordinary accomplishments of the neuro-sciences over the past twenty-five years and their assimilation by philoso-phers in the field of consciousness studies. Although there might be de-murrals about particular points here and there, the current picture is clear enough. The brain involves somewhere between 50 and 150 billion neu-rons; let's say 100. These are a variety of fine, threadlike, long "brain cells" that are not only wound up inside the brain but that extend through-out the body to link to your brain everything from your big toe and five senses to your internal organs. Within the brain these neurons connect with each other via synapses across which neural impulses send electro-chemical "messages." The sheer number of connections is beyond reck-oning, greater, it is said, than the number of stars in the universe. Besides registering the performance of the body, this network is the place where cerebration, emotion, and all forms of psychological experience take place. The sheer activity going on every microsecond means that our sense of the smooth continuity of our consciousness is a gross illusion, like the illusion of visual continuity. In the case of our eyes, 100 million rods and 7 million cones in our retinas—the receptors of light from the scenes we behold—send electrochemical impulses to the brain via neu-rons. Since our range of clear sight consists of a very small area directly in front of us, we are constantly refocusing our eyes and moving our head at the rate of several "saccades" (or eye movements) per second. This means that the smooth-seeming panorama that we view is completely re-drawn several times a second, since every rod and cone receives a differ-ent light particle with each refocus. Just as we don't hear the forty-four thousand interruptions per second between the samples of music coded on a compact disk, don't see the individual frames of a movie film or the many redrawings per second of our TV and computer monitors, we are completely unaware of the pointillistic nature of our vision.

In the case of consciousness, an entirely new round of neuronal firings takes place every microsecond, some consisting of recursive repeats of sig-nals already sent (producing stronger and more lingering effects) while most of the others occur so briefly that they are almost immediately lost. The information contained in these firings derives from the total system of our being, both bodily and mental, an influx again beyond reckoning. "You can't cross the same river twice" is herein given radically new mean-ing. And if you don't believe it, consider what happens when you've typed a long email outpouring to a friend only to hit a wrong key accidentally or suffer a two-second power failure. The email is lost in cyberspace. Can you write it again? There is no chance in the world that you can recreate that letter exactly as it was. Except for the most salient points, the rewrite will be a vastly different letter. You're just not the person you were a few seconds ago; "you" have been redrawn. Despite all the continuity of what you take to be your "self," there is no stable entity existing behind all the neuronal flux of your brain. Call it a soul, a spirit, a self: it just isn't there. Or as Dennett likes to say, there's nobody home.

One of Dennett's most striking satirical metaphors is the Cartesian Theater, an imaginary place in the head where a spirit or spook or self, usually referred to as a homunculus, watches the movies or stream of consciousness flowing through the brain. Descartes had centered it all in the pineal gland situated in the midbrain, but the concept of an internal viewer was doomed from the start by the infinite regress it entails, since where does the homunculus get his information from amidst the "pandemonium" (a favorite philosophic term to describe the frantic signalizings of the neurons), unless a prior viewer or organizer/indexer has already gotten it all together? One then needs an infinite number of prior viewers to explain each viewer's knowledge.

In reality, as Dennett and most other neuroscience thinkers see it, there can be no center in a pandemonial system of a hundred billion neurons firing whole universes of signals. Since for Dennett there is no center where all their *parallel* inputs are sifted into a *serial* stream to produce a coherent narrative assembled like a Jane Austen novel, he early on developed his theory of "Multiple Drafts," in which signals vie like sperm cells to get through to consciousness, with the winners serendipitously falling into a serial line that has come to be known as a stream of consciousness. No Central Intelligence Agency, so to speak, has selected or arranged them to produce a coherent story. This stream, assaulted on all sides by its rivals, is vulnerable to confusion, forgetting, self-contradiction, and whatnot. (It's amazing that we have as many periods of lucidity as we do.) In *Sweet Dreams,* Dennett updates Multiple Drafts into "Fame in the Brain," a new metaphor to describe the victorious emergences from pandemonium into consciousness, which he likens to the vagaries and happenstances of celebrity (e.g., Rodney King).

Since without an organizing center the self is purely virtual, there is no

"I" to make the choices that could give any meaning to the expression "free will" (the "I" would have to be a homunculus that selects which firings get through, leading back to the infinite regress problem). This in turn leads to the big question, nowadays called the "Hard Problem" (as opposed to the "easy" neuronal one answerable from MRIs of the brain and other forms of reporting) of what exactly *is* consciousness if there's nobody home.

Enter "zombies."

Sweet Dreams updates Dennett's decades-long talk about zombies by introducing another one of his characteristic epithets, the "Zombic Hunch." And what is a zombic hunch or, for that matter, what is a zombie? A zombie is a philosophic term to refer to an imaginary creature just like us in every way except for its lack of consciousness. Why is such a creature even thinkable? Because, to refer to the epigraph above from Dennett, "Only a theory that explained conscious events in terms of unconscious events could explain consciousness at all."[2] The unconscious events that lead to our consciousness are the zillions of firings already discussed. (After all, life is produced from nonlife, unless you're hopelessly addicted to magic spooks. And how do you explain *them*?) If our consciousness is produced from multitudes of unconscious firings, then it is plausible to Dennett that a computer robot could eventually be produced (from multitudes of bytes via silicon chips that substitute for neurons) that could mimic real people without having to be conscious. "How can a little box on your desk, whose parts know nothing at all about chess, beat you at chess with such stunning reliability?"[3] Dennett asks. Or to put it even more succinctly: it does not require consciousness to produce consciousness. (Or else you would fall into another infinite regress.)

"There is a powerful and ubiquitous intuition that computational, mechanistic models of consciousness, of the sort we naturalists favor, *must leave something out*." But what is being left out? Nobody can quite say, and yet they insist *"that there is a real difference between a conscious person and a perfect zombie*—let's call that intuition the *Zombic Hunch"* (SD, 13–14). But when you seriously think about it, when you realize the sheer involuntariness of our behavior and thoughts generated out of the whole pandemonium of neuronal activity that produces us, it does indeed become hard to say what is left out. If one compares a "perfect zombie" with an actual person whose consciousness has been produced by activity that is not conscious (as "life" has been produced from "mud") what's been left out, presumably, is "consciousness" itself. But what, to echo Virginia Woolf, is the thing that lies behind the semblance of the thing? Now, alas, we're going around in circles.

In the closing pages of *Consciousness Explained,* Dennett sums things up:

> The phenomena of human consciousness have been explained in the preceding chapters in terms of the operations of a "virtual machine," a sort of evolved (and evolving) computer program that shapes the activities of the brain. There is no Cartesian Theater; there are just Multiple Drafts [now Fame in the Brain] composed by processes of content fixation . . . It is indeed mind-bogglingly difficult to imagine how the computer-brain of a robot could support consciousness. How could a slew of information-processing events in a bunch of silicon chips amount to conscious experiences? But it's just as difficult to imagine how an organic human brain could support consciousness. How could a complicated slew of electrochemical interactions between billions of neurons amount to conscious experiences?
> (CE, 431, 433)

By the time we reach *Sweet Dreams,* another ongoing Dennettian concept reappears once again: heterophenomenology. This literally means the phenomenology—or conscious experiences—of another person, or what we would call a "third person," as opposed to ourselves, the "first person." Since it's our own phenomenology that is the big question (our consciousness and experiences) and since there is no way an investigator can directly inspect another's first person reality—in other words, his subjectivity—which exists only for the "first person" himself, how can we hope to explore what first personhood, the essence of consciousness, consists of? (What is the thing beneath the semblance of the thing?) For Dennett, heterophenomenology, examining other people's reports about their consciousness, observing their behavior, is good enough. Furthermore, if consciousness is ultimately composed of material particles (electrochemical firings) it ought to be as accessible directly and indirectly as the elementary particles of the physical sciences that we have inferred so much about from various types of signs.

In sum, for Dennett there is nothing extra or left out, no special material of consciousness over and above what we can infer from other people's behavior and reports, especially when added to the increasingly sensitive "photographs" being taken of the brain in action. For Dennett, "consciousness" is a stubborn spook that some of us refuse to let go of. Just as the vitalists believed that life was something added to matter—until the sciences demonstrated otherwise—there are "mysterians" who insist that consciousness is something over and above matter. (But if it's not matter, what is it if the ghost in the machine has presumably been banished?) "A hundred years from now, I expect this claim [of something extra] will be scarcely credible" (*SD*, 14), is Dennett's prophetic response.

In a famous essay, Thomas Nagel said that we can never know what it feels like to be a bat. Unsurprisingly, Dennett is not sympathetic to this stance, since he believes that everything essential to this knowledge can be made accessible through third-person examination, even if bats can't talk. To support this, he has long accounts in his writings of yet another imaginary being who has become a staple of philosophers of consciousness, a person he variously calls Robo-Mary or Mary-Mary. Mary, a brilliant scientist-scholar, lives in a room without colors, entirely black and white. Her own body has been camouflaged to hide its colors, and her TV and computer monitors are black and white. She has studied every aspect of color: light, frequencies, reflection, photons, you name it. There is nothing physical about colors that she does not know. "Mary had figured out, using her vast knowledge of color science, *exactly what it would be like for her to see* something red, something yellow, something blue in advance of having those experiences" (*SD*, 106). And so the various philosophers who tell this story finally ask: When Mary eventually emerges from her protected cell and sees a yellow banana, will she undergo an experience that she has never had before, despite all her knowledge? Dennett says no.

On this matter, my impression is that he is quite in the minority. If we turn to neuroscientists (as well as other philosophers that Dennett sees as "reactionaries") we find responses like this:

A conscious human being and a photodiode can behave similarly, at least under certain circumstances. They both can differentiate between light and dark. We know how the photodiode does it. We also know reasonably well how a human being can do it, since we know there are neurons in the retina and in the visual areas of the brain that fire differently, depending on the amount of light. But why should the human's, but not the photodiode's, ability to differentiate between light and dark be associated with a conscious experience of light or dark? Why should the firing of certain neurons in the human visual system generate a "quale" of light, a subjective feeling that there is light, while the voltage change in the photodiode does not?

The authors, Gerald Edelman and Giulio Tononi, conclude "that consciousness is embodied uniquely and privately in each individual; that *no description, scientific or otherwise, is equivalent to the experience of individual embodiment*" (italics mine).[4] I have carefully avoided introducing the philosophical term/concept "qualia," a hornet's nest threatening many needless stings. Edelman and Tononi's "subjective feeling" will serve for present purposes.

Although Dennett has some allies in his view that consciousness is not something over and above the electrochemical firings of the brain (unless you want to start messing around with spooks again), he also has plenty of opposition. Most notable is David Chalmers, author of *The Conscious Mind: In Search of a Fundamental Theory* and many papers and articles.[5] Chalmers, like Dennett, is largely a "materialist," but by no means entirely so, because he does not believe that materialism can answer the question "What is consciousness?" even if it can pretty well answer everything else. ("Materialism" as a concept in philosophy entails the belief that everything can be explained in terms of the physical sciences, without recourse to spooks.) Thus he—and especially Dennett in attacking him—speaks of himself as a dualist, risking the ignominy of that stance in a post-Cartesian world. But he adds, "It is an innocent version of dualism, entirely compatible with the scientific view of the world. Nothing in this approach contradicts anything in physical theory" (*JCS*, 210).

Chalmers's book, and more effectively his article, "Facing Up to the Problem of Consciousness," are mainly concerned with solving the problem of what consciousness adds to a purely materialist version of reality. In attempting to do so, Chalmers appropriates an everyday word: "The really hard problem of consciousness is the problem of *experience*. When we think and perceive, there is a whir of information-processing, but there is also a subjective aspect. As Nagel (1974) has put it, there is *something it is like* to be a conscious organism" (*JCS*, 201) This subjective aspect, the felt quality of experience, is what Dennett has reduced out of, distilled from, serious knowledge, instead explaining everything by means of third-person reportability. At the end of books like *Consciousness Explained*, Chalmers believes, "the author declares that consciousness has turned out to be tractable after all, but the reader is left feeling like a victim of a bait-and-switch. The hard problem remains untouched" (*JCS*, 202). That hard problem is experience, "the most central and manifest aspect of our mental lives."

Is "experience" just another spook whisked in from a back door? "Experience may *arise* from the physical, but it is not *entailed* by the physical . . . In particular, a nonreductive theory of experience will specify basic principles telling us how experience depends on physical features of the world. These *psychophysical* principles will not interfere with physical laws . . . Rather, they will be a supplement to a physical theory" (*JCS*, 208, 210). Chalmers describes his outlook as a "naturalistic dualism," from which vantage point experience would be regarded "*as a fundamental feature of the world, alongside mass, charge, and space-time. If we take experience as fundamental, then we can go about the business of constructing a theory of experience*" (*JCS*, 210; italics mine).

Chalmers shares the powerful intuition of most reflective human beings that there is something about consciousness over and above data processing, a position that Dennett well understands and for which he has a residual sympathy mixed with contempt: "It seems to many people that consciousness is a mystery, the most wonderful magic show imaginable, an unending series of special effects that defy explanation. I think that they are mistaken: consciousness is a physical, biological phenomenon—like metabolism or reproduction" (*SD*, 57). Or to put it most strongly and conclusively, "It is a very good bet that the true materialist theory of consciousness will be highly counterintuitive (like the Copernican theory—at least at first)," which means that traditional philosophy will remain hung up on its "conservative conceptual anthropology until the advance of science puts it out of its misery" (*SD*, 129).

At this juncture, I need to take the risk—as an educated layman—of pronouncing judgment on these rival "solutions" to the maddening problem of consciousness, a problem far from being simply academic and that seems as important as the deciphering of the physical universe in telling us who we are.

I've been reading Daniel Dennett for several years now, feeling mostly edified by and acquiescent to just about everything he has to say. His neuroscience, his demolition of self and soul, his philosophical materialism and anti-dualism, his dismissal of "free will," his promulgations of Darwinian evolution, and his blurring of the distinction between philosophical zombies and human beings all strike me as illuminating, even as I acknowledge that zombies lack "inner" understanding (but "inner" may just be a deeper level of "outer"). Indeed, in a recent essay of my own on the relation of consciousness to creativity and "free will," the penultimate sentence reads: "If we're robots, we're pretty damned brilliant robots."[6] It's a fairly accepting "if," not meant to suggest doubts about the neuronal network that motivates us but unwilling to deny the uniqueness of human understanding.

So when Dennett claims he has explained consciousness, or that a computer program will someday simulate consciousness, I'm dissatisfied with a solution that essentially solves the problem by eliminating it altogether. The obsession that there is "something extra," which he claims science will eventually put out of its misery, refuses to go away. It's the feeling of being conscious (of experiencing *qualia*), which Dennett dismisses as an intuition as unfounded as the belief that the sun moves around the earth or that one's soul has been installed by a "creator" at birth. I agree that evidence of qualia cannot be physically found anywhere and that qualia are like the Cheshire cat's smile with the cat removed, but qualitative existence is felt nonetheless. (Will a computer program ever write "real" poetry—or just poetry that sounds like John Ashbery?) Dennett has walked a brilliant walk but finally is stopped by the brick wall of consciousness, even though he makes like someone walking right through it.

David Chalmers, on the other hand, while sharing a majority of Dennett's insights, doesn't buy the claim that consciousness has been explained. For him (as for me), it has simply been denied or evaded by being subsumed into the operation of neurons. It *feels* wrong, and for Chalmers, *experience* is the concept (and reality) that has been neglected. But how successful has Chalmers been in defending, or substantiating, the reality of experience? His magnum opus of 1996, *The Conscious Mind*, consists of 414 packed pages (including the index). It starts out well, describing his dissatisfaction with a purely materialist treatment of consciousness, insisting that the feel of it is not something extra or supererogatory but is part of personhood itself (and to some degree animalhood). But things start to go downhill somewhat quickly thereafter. Pages and pages are devoted to densely peripheral byways of obscure philosopherese, so peripheral in fact that Chalmers has put asterisks at the heads of these sections to indicate that they could very well be skipped! Good advice, since these passages lead nowhere. When he finally pulls out his plum of "experience," we perk up, nod yes, and hope for the best. Alas, a forlorn hope. Experience is not a spook but on a par with mass, charge, and space-time? One gulps. After that, the descent is rapid. On page 277 we read:

> For a final theory, we need a set of psychophysical laws analogous to fundamental laws in physics . . . When combined with the physical facts about a system, they should enable us to perfectly predict the phenomenal facts about the system . . . This is a tall order, and we will not achieve it anytime soon. But we can at least move in this direction . . . The ideas in this chapter are much sketchier and more speculative than those elsewhere in the book, and they raise as many questions as they answer. *They are also the most likely to be entirely wrong* [italics mine].

The center, it seems, will not hold. Things fall apart. I'm left with the Really Really Hard Problem: accepting as I do Dennett's view that there is nothing to consciousness that is not physical, disbelieving in any sort of spooky extras, liking the idea of "experience" but suspicious of its veering into the *je ne sais quoi,* I am willing to entertain that this may never be solved. Yes, I agree with John Searle and others that hydrogen and oxygen in certain proportions metamorphose from gases into something totally different: water, a slippery liquid. And maybe, in a sense, physically generated particles can somehow metamorphose into first-person reality. But once again: What is a thought? What is a thinker? What is the thing that lies behind the semblance of the thing? I doubt that I will be around to learn the answer.

The Crumbling Mortar of
Social Construction

Although Steven Pinker would seem to have done a definitive job de-molishing the quaint notion of the human mind as a blank slate upon which anything may be written, this strange fancy persists in the hu-manities and elsewhere. Witness the statement by Robert Scholes, former president of the Modern Language Association: "Yes, we were natural for eons before we were cultural—before we were human even—but so what? We are cultural now, and culture is the domain of the humanities."[1] At this late date in our knowledge of human cognition, affirming blank-slatism seems little more than a will to power, a triumph of ideology over probity, a politically correct disdain for truth. If human nature is deter-mined by culture, rather than the reverse, then all things are possible, from Holy Ghosts to ghosts in the machine, just about anything one's heart's desire is capable of inventing. It was not so long ago when even the intelligentsia refused to believe that *Homo sapiens* was an offshoot of apes. Though it walked like an ape, ate like an ape, produced excrements like an ape, had sex like an ape, was gestated and born like an ape, had the same body parts as an ape, *looked like an ape,* and finally died like an ape, it was in reality the image of God (who presumably did not "him-self" resemble an ape).

When we sense the blank-slatism that underlies creationism and intel-ligent design, we have no trouble brushing it off as a psychological aber-ration, a form of denial with little regard for truth. But when we hear it from professors and their graduate students—and the head of the MLA—it's alarming. If "God" as ventriloquized by the pious is really a naked power ploy to dictate culture by making up stories about what "God" in-tends, social constructionism is only a thinly veiled version of that ploy. "Culture" is substituted for God and ventriloquized by academics who fancy, in bursts of narcissistic *fiat lux,* they are creating that culture through scholarly articles and books. This misbegotten "creativity" decides in advance that certain conclusions—about race, intelligence, evolution, free will, and so forth—are out of bounds. Truth? In his presidential farewell address to the Modern Language Association in December 2005, Scholes alluded to the falling star of the humanities, quoting a remark by George Steiner to the effect that in the sciences there is an obligation to accuracy and truth based on real knowledge, but in the humanities, you can say just about anything. "What is truth?" asked jesting Pilate, but he didn't wait for a reply. Neither did Scholes, who simply dropped it and went on to other concerns. When I wrote a letter of protest, Scholes responded with that blast of cultural triumphalism. Nature? Forget about it!

Let's shift gears and consider a hypothetical middle-aged man we'll call Hosa, for *Homo sapiens.* When Hosa arises in the morning, usually logy and drooped in spirits, especially if the sun is not already shining, he badly needs a powerful caffeine fix. Since he's a professional of some sort, he spends some days at home or on the road and others at his office. This morning, he retrieves the newspapers from the driveway, settles into a chair, and awaits the bolstered animal spirits usually produced by the coffee, which doesn't take very long. His psychological state gradually improves, but after two large mugs of brew he starts feeling a vague discomfort distracting him from his reading, which suddenly becomes a need to urinate. Once relieved, he returns to a clearer focus on reading until, again vaguely uncomfortable, he realizes that the customary early-morning sexual restlessness from testosterone levels that peak around 4 a.m. has been itching his body and mind for hours, stealing that focus away with sexual images and fantasies. If this had been a day at the office, the itch would disappear as he raced off to work, but being at home means that in one way or another it has to be scratched. However satisfied, this opens a time-consuming scenario that may perhaps produce relief but only serves to intensify growing hunger from lack of breakfast.

Food replenishes a failing energy supply—an energy supply that drives bodies and brains the way gasoline "drives" a car—but leads to drowsiness and dulled wits if the proteins are low and the carbs are high. Siestas, after all, were not "constructed" out of whole cloth but sprang from biology, postprandial slump, and hot weather, even if in most of the Western world there usually isn't any time for them. All day long, the chemistry of the body is charging and draining, like a car battery, but also charging and draining Hosa's psyche and changing his point of view from minute to minute. Sometimes he's impatient and irritable, angry at his wife, his neighbors, his professional associates, or his congresspersons; at others he's amorous and sensual, finding some people attractive he wouldn't otherwise look at twice; at still other times he has unforeseen brilliant insights. Alcohol causes his writing to be sloppy and his tongue to be a bit loose to his colleagues. Air pollution,[2] sunlight, barometric pressure, a deficiency or superfluity of glucose, vitamins, potassium, whip his body and mind through one chemically induced state after another. You don't have to be a druggie to be a druggie.

These transient states, the unwilled streams of consciousness, the invisible teleprompter from which we unwittingly read our scripts in what Daniel Dennett jokingly called the Cartesian Theater—these are the normal sustaining features of skittish everyday sentient life, but Hosa is rarely aware of them as such, however much the most transient microburst of an image can redirect his emotions and thoughts. At every heartbeat, this volatility is not only operating but could be easily visible if he were to pay attention—but the usual focus of attention is only upon the individual states, not their fast-moving alterations. Although he would certainly notice the effects of spinach laced with *E. coli,* would he recognize that he is no longer attending to his reading because a face in a photo has revived the bitterness of yesterday's put-down from the head of his department? Given the volatility of this mental stream, is there ever a neutral point of the day in which Hosa is in godlike stasis, at his best no matter what the task to be accomplished? A time when drowsiness, the need to excrete, libidinousness, hunger, inebriation, elation, irritation, depression, are not being set off by high or low humidity, circadian rhythms, heat or cold, transient thoughts, too many refined carbs, insufficient protein, a current of toxic air from a distant copper smelter? A time when he is simply a clear-thinking spirit uncontaminated by the body and its moods? Is there a default mode of pure thinking bodilessness set free from biochemistry?

In other words, are there times when great things are said and done *despite* the "contamination" of thought by the body? But how could such moments take place "despite" the body, when thoughts are produced *because* of and *from* the body? One is always in a mood and all moods are produced by biochemistry in conjunction with neuroanatomy. Thoughts, like pearls, are the products of somatic irritation. "I think therefore I am" should be replaced with "I feel therefore I think." It's not "Cogito ergo sum." It's "I metabolize, thereby I exist in moods of thought." Since I am always in a mood, validity can only mean that I am speaking to other people who also have moods and that some of them are more receptive to my thoughts than others. Widespread "moods" have come to be called "human universals." The arts wouldn't exist without them. Politics depends on them. Society is constructed from them. And truth? Some moods produce stunning insights into reality, not an in-itself reality, of course, but a reality intelligible to a knowing human nature because of the brain's inbuilt structures of understanding. A knowledge of things in themselves is a logical contradiction, because things in themselves do not have a human brain "knowing" them by means of conventional categories. Even mathematical truths are sequenced in human time via equations attuned to brain capacity. God would know it all at once.

Since I have already taken the liberty of imagining a typical Western *Homo sapiens,* I might as well go further and imagine a virtual self—ongoing but changing every minute with its seemingly aleatory biochemistry. If the stream of consciousness is not produced by a freestanding spook outside of the neuronal system, the notion of an ongoing virtual self is nevertheless no more difficult to imagine than the notion of an ongoing physical person: just about all the material components of our bodies are replaced or altered over time, yet "we" (as bodies) remain recognizable to ourselves and other people. This virtual self, in fact, is the only self there is. It's the real self, projected from billions of synapses of a hundred million neurons. Its particular character is the result of millions of years of evolution, the individual's inheritance of genes, their actions on the internal biochemistry of his or her particular brain, and the inputs from society, sometimes referred to as cultural constructions.

With this in mind, let's do a little thought experiment: Hosa wakes up one morning and no longer experiences hunger but instead finds a bottle of pills that must be taken a few times a day with lots of water. This becomes the new fuel, a miraculous fiber-and-bulk-free lifegiver. What will happen to Hosa's day, to his life?

An enormous part of Hosa's day will become vacated, as well as an enormous component of his consciousness and an enormous alteration of

biochemically induced moods. The time slots formerly allotted to breakfast, lunch, and dinner will now be reduced to a few minutes of pill-
popping. The breakfast and dinner tables will no longer be needed, nor
their chairs. The dishes on which so much time was spent to select, purchase, set out on the table, fill with food, put into the sink and dishwasher,
and break, will no longer be needed. The trips to the supermarket and the
natural-foods stores will disappear as well as all the walking or driving
involved to get there. Wear and tear on the car will decrease as well as gas
consumption. All the snacks and chips and junk food will no longer be
a problem, nor will obesity (maybe something else will take its place)
and maybe not even diabetes type 2. Hosa's whole financial situation will
change, since no money will have to be spent on food, wine, beer, or
restaurants. Without dinner parties and restaurants, his social life will undergo considerable alteration. Meeting his pals at the local bar won't be
meaningful, nor will going on pig-out ocean cruises have much appeal.
Stains on his clothing from spilled food will be a thing of the past, so dry-
cleaning and washing bills will decrease. Food fights with his children will
stop, since it won't be necessary to do more than force a few pills down
their resisting throats. Many hours of the day will now be vacant for
other things, both good and bad. Diet-related diseases will disappear. The
psychological landscape will change, since psychological spikes and defi-
cits from nutrients will stop. Not taking one's pills will be a new source
of troubles. Arranging to have freshly baked bread come out of the oven
when potential buyers arrive to inspect his for-sale house won't be needed
anymore, because the smell of food will be meaningless. We can begin to
see that what looks at first to be purely personal alterations of Hosa's life
gradually takes on powerful social dimensions.

From society's point of view, the changes will be drastic. There are now
all sorts of things that society will no longer need to "construct." Dishes
are the least of it. The industries that manufacture them will disappear,
along with the industries that produce pots and pans, dinner napkins,
cooktops, ovens, microwaves, freezers, and refrigerators. All sorts of
furniture will no longer be needed, not to mention table linens, corkscrews,
food shopping bags, food cartons, cans and their openers. Multiplex the-
aters will probably experience financial problems without the huge quantities of overpriced junk foods they sell. With Braeburn apples no longer
flown from New Zealand to the USA, jet planes will have less cargo, not

to mention the whole trucking industry that brings fresh produce from Florida and California. The wear and tear on the interstate highway system will decrease, affecting a wide range of industries. America's relationship with Middle Eastern oil empires will doubtless change, since we won't be needing all that petroleum. The restaurant industry would vanish, along with thousands upon thousands of Starbucks and McDonalds. Strip malls would change in character. *Wine Spectator* and *Gourmet*, along with many other magazines, would have no subject; *Organic Gardening* would discuss only flower gardens. Vast areas of land now devoted to farming would have to be given other uses, maybe as national parks or ethanol production sites. The advertising and newspaper industries would suffer a huge blow, since the Wednesday food supplements would no longer be published.

All of the disappearing elements of current daily life could be referred to as "culturally constructed." But only in the most literal bricks-and-mortar way. The fashionable academic sense of "invented in order to populate a blank slate" (which implies that ideology and not biology is what makes and remakes society) would now be meaningless, since even the most politically correct assistant professor would be forced to concede that the slate, far from being blank, was indeed all along the source and impetus for the bricks-and-mortar constructions of society. For how could any society construct customs or machinery to express and satisfy something they knew nothing about (i.e., something not motivated by the human biological program)?

If we were to use sex as our test case, the results would be even more shocking. After all, *Homo sapiens* is the only primate with concealed ovulation and all-year-round sexiness and receptivity. What if we were suddenly to find ourselves like apes and monkeys with a brief mating season once or twice a year? These seasons are triggered by the interactions of the environment with internal hormonal clocks and neuronal structures, manifesting themselves in gross, visible signs of receptivity, such as changes in the appearance of the female anal perineum and suddenly attractive biological smells. Without 24/7 sexuality, a list of what would disappear would take up more pages than the contents of this essay. Cosmetics, sexy clothing, romantic love, *The Mill on the Floss*, TV dramas and soap operas, flirting, *Sex and the City*, MTV and music videos, condoms and birth control pills, quickie hotel rooms, sexual politics of today's sort (a new sexual politics would doubtless evolve), genital mutilation, "rape" (it would become trickier to distinguish rape from normal everyday sex)—the world of today would be turned upside down. Clothing manufacturers would be even more concerned than they are now with male peacock fashions that would influence the sexual preference of females during their brief fertile periods. And according to an article in *New Scientist* entitled "Why the Best-Dressed Women Have Babies on Their Mind," even now "women take greater care over their appearance when they are at peak levels of fertility."[3] More pressing than Columbus Day sales would be Copulation Day sales. (After which you and your sweetie may both lose interest in sex altogether for another six months or year.)

Culture is slave to body and brain, not their master! It "constructs" bicycles for bipeds, not quadracycles for quadrupeds, since there aren't any hominid quadrupeds. Indeed, it constructs entire worlds, known as societies. But they are worlds wholly at the service of the evolved body and its fleeting moods. The body arrives with millions of years of baggage from selection and adaptation. Up to a point the bodies and psyches can be remarkably altered by culture, as we see from today's cyborgian prostheses and the weird rites of religious cults and the influential lunacies of a Britney Spears. But the alterations are always shaped and limited by the human nature they are addressing. This human nature can very well change over time through natural selection. But culture is always at its service, no matter how bizarre the expression.

The anti-science attitudes of humanists are one more face of ideology-driven politics. Ideologues prefer to preen themselves with their supposed power over blank slates, which gives them an hubristic sense of illimitable power to construct people as they wish. But they often end up like Chance the gardener in the film *Being There*. Chance was unable to distinguish between the grandiose power he could exert over TV by clicking his remote control and his impotence to alter real life no matter how furiously he clicked. Moreover, ideologies and beliefs can only shape what's there, not what's not there. What's there are bodies and brains with natures of their own, brilliantly flexible but always vulnerable to extremes to which they finally must say no! And if they stupidly fail to say no!, species-specific Nature will inevitably do it for them.[4]

My Life as a Robot

"FROM ENVIRONMENTALISM TO CONSCIOUSNESS"? This book has attempted to explain the connection. My awareness of the effects of the environment on body and consciousness came about as I began to understand how toxic substances, pollution, the quality of soil in which food is grown, living near highways, chlordane, PCBs, DDT, global warming, lead in paint and dishes, and so forth are more than casually related to one's physical and mental condition. Did the fact that my mother smoked when I was gestated (and that we daily breathed the sooty particulates of New York City life) play a role in my frailty as a child, my great susceptibility to serious childhood diseases, my rheumatic fever and heart disease (although I eventually overcame the worst consequences)? What effect does childhood infirmity have, not only on one's body but on one's psychological persona for the rest of one's life? It can't be negligible.

Ecological awareness, when informed by an understanding of evolutionary biology, can only serve to further constrict any fantasies one might have about human "freedom." After all, what can "freedom" mean? Having five fingers on each hand rather than three, having toes and not webbed feet, walking upright, etc., etc.—each and every physical characteristic genetically amassed over millennia into an illusory, transient "normality" known as *Homo sapiens* results in a certain limited bodily construction and range of activity. (If everyone were color-blind, inability to distinguish red from green would be "normal.") With this small and limited set of "normal" characteristics all other possibilities for existing in the world are, at least temporarily, ruled out. "Normality" is not an essence or a transcendent fact but (from the point of view of "freedom") a constricting happenstance of the environment, natural selection, and genetics. Had the environment of the planet been slightly different at a certain moment of the past, a different range of survivors would have evolved to produce us with a different normality. Human nature or "normality" are pure contingency, but these contingencies rule the range of what is possible for us.

Darwinian evolution and behavioral ecology teach powerful lessons about "free choice" and selfhood well beyond the physical construction of our bodies. They carry the message of environmentalism into more complex territory. But neuro and cognitive sciences go far beyond hearts, kidneys, feet, and endocrine systems: in the late twentieth century these sciences have led to the shocking awareness that a network of billions of neurons and millions of billions of synapses are a self-directing system that produces *us* as already-mapped-out but seemingly flexible psychological beings bemused by the sometimes tragic-comic illusion of autonomy. At this very moment, as I sit at the computer and write, I am unable to locate the very same "I" that David Hume failed to locate several hundred years ago. "For my part [to quote once again one of the most marvelous passages in British philosophy], when I enter most intimately into what I call *myself,* I always stumble on some particular perception or other, of heat or cold, light or shade, love or hatred, pain or pleasure. I never can catch *myself* at any time without a perception, and never can observe anything but the perception." A stream of thoughts races through my head and I dutifully write down these unchosen ideas.

Indeed, since early this morning, as I arose from bed, I have been hearing Bach's motet *Der Geist hilft unser Schwachheit auf* ("The Spirit doth our weakness help") in my head. It is not as though I have been playing this music from a compact disk—the music has been playing me. The last time I placed disk in player to hear it was months ago when I was working on a project about Bach in the twenty-first century. During that period, I had listened more than twice over to all of Bach's 215 or so surviving cantatas and all the major choral works (which I could probably sing in my sleep—and doubtless do). Why is this particular composition playing me today, seemingly out of nowhere, just like the thoughts I am transcribing from my internal teleprompter, though without any real consciousness of choosing to do so as a me? Is there any activity or state of consciousness during the entire day or night that such a me has chosen? The essential difference between dreaming at night and thinking during the day is not the voluntariness of one and the passivity of the other: it's only the apparent coherence of one versus the quixotic surreality of the other.

All of it is passivity. With no "I" as conductor of the symphony, the music of the neurons plays out however they please.

Should I therefore begin to feel troubled about the nature of my so-called "spiritual life"? What is a spiritual life anyhow and do I really have one? It *sounds* like some higher state of consciousness, a testimony to my exalted nature as an image of God. But if no "I" has put it all together, if it's not inspired or directed by some transcendent force, if it's in no way different from dreaming or thinking, what on earth does my spiritual life consist of that is missing from my grosser cerebrations? Especially if there is no such thing as "spirit." (Bach was overly optimistic.) Consider how easy it is to manufacture entities that have no existence independent of my brain. In the case of Bach, for example, we have inherited the distinction between his "sacred" cantatas and choral music and his "secular" cantatas. What does their difference consist of? Is there a "sacred" (i.e., "spiritual") music whose *prima materia* is different from vulgarly "secular" music? The answer is unambiguous: it's simply no! Bach's most "sacred" music—the *Mass in B minor,* the *Christmas Oratorio*—was largely composed by recycling music from the secular cantatas, much of it reused almost note for note. (The *St. Matthew Passion* is an exception to this practice.) The difference between them is the religious texts. Do priests become sacralized when they don their outlandish costumes, when they quote the "word of God," or in the final analysis are they only variants of the Wizard of Oz? Spiritual and secular turn out to be yet another phantasmal dualism that cannot be substantiated in any way. And religion? A different essence from everything else—or just politics with a pious gloss? A form of moral blackmail pretending to "higher" and more "spiritual" powers?

Daniel Dennett, you will recall, did not see any substantive difference between computer chips and neurons. "It is indeed mind-bogglingly difficult," he wrote, "to imagine how the computer-brain of a robot could support consciousness. How could a slew of information-processing events in a bunch of silicon chips amount to conscious experiences? But it's just as difficult to imagine how an organic human brain could support consciousness. How could a complicated slew of electrochemical interactions between billions of neurons amount to conscious experiences?"

Despite the fact that just about every chapter of this book would lead a perspicuous reader to infer that I see myself as essentially a "robot," you will also perhaps recall that I ended my treatment of Dennett with this remark: "Accepting as I do Dennett's view that there is nothing to consciousness that is not physical, disbelieving in any sort of spooky extras, liking the idea of 'experience' but suspicious of its veering into the *je ne sais quoi,* I am willing to entertain that this may never be solved . . . Maybe, in a sense, physically generated particles can somehow metamorphose into first-person reality. But once again, what is a thought? What is a thinker? What is the thing that lies behind the semblance of the thing? I doubt that I will be around to learn the answer." This sounds like a certain amount of skepticism about Dennett's conclusions. But as the "author" of the above remarks (What is an author? See chapter 20), I've got to confess to a certain skepticism of my skepticism. I was uneasy when I wrote that conclusion then and am just as uneasy now. But what's to be skeptical about? That spooks don't exist? From ecology I learned that myriad elements of the "environment" are woven into my physical being, influencing how I feel and how I exist. From my evolutionary and biological readings I learned that the very composition of my cells is the result of millions of years of evolutionary trial and error that I happened to survive (or I wouldn't be here writing this now—obviously my multitudinous forebears were a tough breed) and my possibilities and actions are shaped by the very number and arrangement of the fingers of my hands and configuration of my spine. (No knuckle-walking for me!) And then I learned from the neuroscientists and the philosophers of consciousness that my brain is not just a sponge that takes everything in—indeed, that it's no sponge at all. Rather it's an unfathomably complex neural network beyond reckoning, a network that *is* my life, that *is* my consciousness, that *is* the thoughts I think and the feelings I feel. Where does my so-called "spiritual life" fit into all this?

Unless I am planning to pull a treacherous last-minute happy-ending intelligent design sort of deus ex machina out of the hard-won knowledge behind this book (in which case you would be justified in trashing it right now) the conclusion seems inevitable: my "spiritual life" is a pious self-regarding hoax.

So where does all of that leave me (or "me")? Is anything radically changed by such an awareness? To what extent am I really a robot, and if indeed I am a robot, does that mean that conversing with me is a waste of time?

For starters, I don't think I would have the slightest difficulty passing a Turing test—that is, convincing an interlocutor that I am not a computer but an actual "person." In other words, unlike a computer or a robot, there *seems* to be something that goes on inside people that we call "reflection." I "think over" what has been said to me and respond accordingly. I *feel* a consciousness within me. (Daniel Dennett reminds us, however, that we *feel* that the sun goes round the earth.) But that means nonetheless that the material of my reflection can only come from what is already stored in my neurons. It can't come *directly* (i.e., unmediatedly) from "outside" but only insofar as the signals from outside alter what is already stored. If they can, they augment the storehouse so that there are more options for reflection in the next round. Still, in the light of Dennett's argument, it all seems eerily computerlike: stimuli (such as photons) are fed in, giving the processor (or brain) more to work with. But somehow, no matter how this analogy is pounded into us, we continue to have a sense of an I who engages in "spiritual" mental activity. If this I, however, is only reading off the already-decided messages on the internal system-generated teleprompter—since it has no powers of initiating agency and only does what it is told—the case for a really "creative" internal life is badly damaged. Or let me put this in a novel way: if Bach was "free," how come he wasn't able to write music that sounds like Prokofiev (whereas Prokofiev was able to write music that echoed Haydn and Mozart, familiar to music lovers as his *Classical Symphony*)?

As I now attempt to slip out of this self-created noose (it's not very tight, I can hardly feel it), as I exit the issues just raised by summing them up as a statement of my skepticism of my skepticism about being a mere robot, I do so by shifting to a new question: what difference can it make to human life if we become aware that we are in fact robots? My answer is that if it makes any difference at all, such an awareness is probably for the better, not for the worse. For if we really value truth over feel-goodism and if our nature is indeed robotlike, we are already a step ahead by acknowledging this. Next, what limitation does such knowledge impose on us? It can't be that it takes away our "freedom," because it's hardly been a secret that "free will" is a self-contradictory nonsense idea. There's nothing to take away. If "freedom" means the freedom to be an unmotivated vacuous airhead, "I" don't want any part of it.

Being a robot doesn't mean that my reflections will lose their value, that I won't be able to understand what people are saying, that I won't be able to feel their pain (or joy), that my sex life will be harmed (sex has always seemed pretty robotic from the start, and one could hardly find a human experience that makes one feel less "free"), and it certainly won't mean that I am "trapped" in a pattern with no hope of change or growth. I'm changing and growing all the time, at least until Alzheimer's hits, which will definitely be a change for the worse (while at the same time providing even more evidence of unfreedom. How much more evidence is needed of electrochemical neuron-dependency?).

If people are not "free," how is it possible that we don't in fact find them locked into a stasis? I have dealt with this to some extent in chapter 20. Certain aspects of our psychological makeups are pretty fixed from the earliest years, our so-called personalities. But experience and rhetoric work surprisingly large changes. As a result of epiphanal or traumatic experiences, we can undergo remarkable transformations. And as a result of rhetoric in the broadest sense, from advertising to the tone of voice and syntax of other people, from the books and periodicals we read, to the impact of films and plays, to the behavior of peers and colleagues, we are constantly undergoing alterations. These changes are not chosen or willed, (there's nobody home, as Dennett keeps reminding us) but they are nonetheless valuable and enlarging. Furthermore, each of us has a powerful influence on other people, an influence than can be honed and refined as our knowledge of human nature and its social expressions increases along with our sensitivity, intelligence, and awareness with regard to human behavior. The drama of human life may not be directed by an all-wise Creator (would a tyrannical Cosmic Legislator make us more "free" than electrochemical neurons?) but ameliorations and enrichments take place anyhow.

Still, a troubling question remains: how can there be a system of morality, law, and punishment if people are not responsible for what they choose or do? Even if the institutions and persons of society are capable of teaching and influencing behavior, the changes that take place are not the result of choice any more than I chose to have my internal music player perform *Der Geist hilft* as I awakened in the morning. There's a clear answer to this that many people will not like at all, preferring their feel-good fantasies ("Jesus loves me even if I'm a sonofabitch!"): failure to punish murderers and terrorists will put an end to civilized life. If someone walks around the streets shooting people, he needs to be incarcerated or worse to stop it, regardless of the impossibility of "free" choices and the unjustness of punishment for behavior that hasn't been chosen. But pretending that it has been chosen is a moral crime of its own. The difference between "by virtue of insanity" and "by virtue of sanity" is nil, except for the useful fact that the "sane" have (involuntarily) learned the lessons society wants to teach because their genetics and their brain development have produced them to be (for the moment) the most-likely-to-survive specimens of "normal human nature," by no means a fixed or transcendent thing. (*If there were real justice in the world, nobody would be given credit for anything.*) The "responsibility," if any, belongs more with society than with individuals, since society is capable of improving its rhetorical skills and its teaching, with immense benefit to its unwitting students. Fascism or totalitarianism are dangerous extremes of regimentation through social teaching, so their fatal social/political vices need to be discouraged by the lessons taught and learned. People have in fact been taught all sorts of beneficial survivalist behavior. With regard to smoking, safe sex, racism, obesity, global warming—changes have been possible, even amidst the atavisms that compromise human systems of morality. But they're not the result of free will.

We tend to treat beautiful people as if they chose to be beautiful (we *like* them for being beautiful, but is there anything more ridiculous than praising them for their beauty?) and ugly people as if they chose to be ugly. We treat brilliant people as if they chose to be brilliant and stupid people as if they chose to be stupid. There may be no way around this evolved and totally unfair behavior; it may be part of our "human nature" and probably serves survivalist ends. But even our human nature has been subject to changes over the millions of years we've evolved. Giving up on the belief in free will could have a softening effect on human attitudes, at best intimating the tragedy of existence, reducing self-righteousness and the blaming of others, an effect that perhaps is all to the good, even if murderers have to be removed from society. Pretending that people have actually chosen is itself a form of cruelty and denial, however satisfying a fantasy. Still, if to understand all is to forgive all, some things will have to remain unforgivable for human societies to survive.

NOTES

Introduction: From Environmentalism to Consciousness

1. See www.lehigh.edu/~inbios/news/evolution.htm.

2. "Reading with Selection in Mind," review of *The Literary Animal: Evolution and the Nature of Narrative,* by Jonathan Gottschall and David Sloan Wilson, eds. *Science* 311 (February 3, 2006): 612–13.

3. See chapters 13 and 15 for accounts of both Alan Sokal's "hoax" and E. O. Wilson's use of "consilience" to reconcile the humanities with the sciences.

4. *Autonomy, Singularity, Creativity,* a project of the National Humanities Center. http://asc.nhc.rtp.nc.us/news/?page_id=4.

Chapter Two: On Being Polluted

1. It's worth adding thirty years later that electronic air cleaners are incapable of removing gases (such as sulfur dioxide), and at best can reduce particulates inside the rooms of one's house.

2. With apologies to Emily Dickinson.

3. Since this was written, throwaway filters capable of catching very small particles are now commonplace for home furnaces and air handlers, replacing the crude woven fiberglass filters that used to be the norm.

Chapter Three: From Transcendence to Obsolescence

1. A. This phrase comes from Wordsworth's sonnet "To Toussaint L'Ouverture" (1802). It was later used as the title of a celebrated book by the classicist Gilbert Highet (1954), but even before that as the title of a book by R. W. Chambers in 1939. I make use of it several times in this book.

Chapter Five: Ecocriticism's Genesis

1. *Iowa Review* 9.1 (Winter 1978): 71–86. More easily accessed in Cheryll Glotfelty and Harold Fromm, eds., *The Ecocriticism Reader: Landmarks in Literary Ecology* (Athens: University of Georgia Press, 1996), 105–23.

2. Michael P. Branch and Scott Slovic, eds., *The ISLE Reader* (Athens: University of Georgia Press, 2003), xvi.

3. Lawrence Buell, *The Future of Environmental Criticism* (Malden, MA: Blackwell Publishing, 2005), 1. Buell's first chapter provides a general overview of the development of ecocriticism. The bibliography is wide-ranging.

Chapter Six: Ecology and Ideology

1. Dave Foreman, *Confessions of an Eco-Warrior* (New York: Harmony Books, 1999); Murray Bookchin, *Remaking Society: Pathways to a Green Future* (Boston: South End Press, 1990); Steve Chase, ed., *Defending the Earth: A Dialogue between Murray Bookchin and Dave Foreman* (Boston: South End Press, 1991).

2. Roderick Nash, *Wilderness and the American Mind,* 3rd ed. (New Haven, CT: Yale University Press, 1982), 129.

3. Bookchin died on July 30, 2006.

Chapter Seven: Aldo Leopold

1. Aldo Leopold, *A Sand County Almanac* (New York: Oxford University Press, 1987), vii.

2. Roderick Frazier Nash, *The Rights of Nature* (Madison: University of Wisconsin Press, 1989), 205.

3. Bill Devall and George Sessions, *Deep Ecology: Living as if Nature Mattered* (Salt Lake City, UT: Gibbs Smith, 1985).

4. Daniel B. Botkin, *Discordant Harmonies: A New Ecology for the Twenty-First Century* (New York: Oxford University Press, 1990), 62.

5. Eugene C. Hargrove and J. Baird Callicott, "Leopold's 'Means and Ends in Wild Life Management,'" *Environmental Ethics* 12 (Winter 1990): 336.

Chapter Eight: Postmodern Ecologizing

1. Lawrence Buell, *The Environmental Imagination: Thoreau, Nature Writing, and the Formation of American Culture* (Cambridge, MA: Harvard University Press, 1995).

Chapter Nine: The "Environment" Is Us

1. Steve Kroll-Smith and H. Hugh Floyd, *Bodies in Protest: Environmental Illness and the Struggle over Medical Knowledge* (New York: New York University Press, 1997).

2. Marian R. Chertow and Daniel C. Esty, eds., *Thinking Ecologically: The Next Generation of Environmental Policy* (New Haven, CT: Yale University Press, 1997).

3. Peter C. van Wyck, *Primitives in the Wilderness: Deep Ecology and the Missing Human Subject* (Albany: State University of New York Press, 1997).

Chapter Eleven: Full Stomach Wilderness and the Suburban Esthetic

1. William Cronon, ed., *Uncommon Ground: Rethinking the Human Place in Nature* (New York: W. W. Norton, 1995), 69–90.

Chapter Twelve: Coetzee's Postmodern Animals

1. J. M. Coetzee, "What Is Realism?" *Salmagundi* 114–15 (Summer 1997): 70–71.

2. J. M. Coetzee, *Doubling the Point: Essays and Interviews,* ed. David Atwell (Cambridge, MA: Harvard University Press, 1992), 52–53.

3. J. M. Coetzee, "What Is Realism?" 83, 101.

4. *Doubling the Point,* 63, 205.

5. J. M. Coetzee, *Life and Times of Michael K* (New York: Viking Press, 1983), 163, 164, 166.

6. J. M. Coetzee, *The Lives of Animals* (Princeton, NJ: Princeton University Press, 1999), 21.

7. Peter Singer, *Animal Liberation,* 2nd ed. (New York: New York Review Books, 1990), 165.

8. J. M. Coetzee, *Disgrace* (New York: Viking, 1999).

Chapter Thirteen: My Science Wars

1. Harold Fromm, "Establishing *a Way* in a World of Conflicts," in *Teaching the Conflicts: Gerald Graff, Curricular Reform, and the Culture Wars,* ed. William E. Cain (New York: Garland Publishing, 1994), 74, 76.

2. *Social Text* 46–47 (This is also identified as vol. 14, nos. 1 and 2, Spring/ Summer 1996). The issue was entitled *Science Wars.* Most of the essays, along with some new ones, were subsequently published as a book in the fall of 1996 by Duke University Press. According to the editors, Alan Sokal's essay was never intended to be included in the book version because of its marginality to the theme of science wars.

3. Paul R. Gross and Norman Levitt, *Higher Superstition: The Academic Left and Its Quarrels with Science* (Baltimore: Johns Hopkins University Press, 1994).

4. See my account of Ross in "Democritus Junior and the Curse of Postmodernism," *Hudson Review* 49 (Summer 1996): 323–30.

Chapter Fourteen: O, Paglia Mia!

1. Camille Paglia, "The MIT Lecture," in *Sex, Art, and American Culture* (New York: Vintage Books, 1992), 265.

2. Paglia, *Sexual Personae: Art and Decadence from Nefertiti to Emily Dickinson* (New York: Vintage Books, 1991), 28.

3. Paglia, *Sex, Art, and American Culture*, 99.

4. Paglia, *Vamps and Tramps* (New York: Vintage Books, 1994).

5. See Paglia's interview in a special issue of *South Atlantic Quarterly: Catholic Lives/Contemporary America* 93 (Summer 1994): 727–46. "People said to me early on, 'Oh, what can you do? One person can't do anything.' And I said, 'Excuse me, one person can move mountains.' That's the example of the saints."

6. I am indebted to John V. Walker for an interesting twist that I derived from his essay, "Seizing Power: Decadence and Transgression in Foucault and Paglia," in *Postmodern Culture* 5, no. 1 (September 1994): http://muse.jhu.edu/journals/postmodern_culture/. I am also extremely grateful for repeated help from Donald Marshall (formerly at University of Illinois at Chicago, now at Pepperdine University) and the late Bill Shuter (Eastern Michigan University).

Chapter Fifteen: A Crucifix for Dracula

1. Edward O. Wilson, *Consilience: The Unity of Knowledge* (New York: Alfred A. Knopf, 1998), 8.

2. John Milton in *Paradise Lost* asks Urania, a goddess of heavenly wisdom, to bring him some inspiration.

3. Wendell Berry, *Life Is a Miracle: An Essay against Modern Superstition* (Washington, D.C.: Counterpoint, 2000), 19.

Chapter Sixteen: The New Darwinism in the Humanities

1. Quoted passim from *The Genealogy of Morals,* trans. Francis Golffing. (New York: Doubleday, 1956).

2. Steven Pinker, *The Blank Slate: The Modern Denial of Human Nature* (New York: Viking, 2002).

3. Jerome H. Barkow, Leda Cosmides, and John Tooby, eds., *The Adapted Mind: Evolutionary Psychology and the Generation of Culture* (New York: Oxford University Press, 1992).

4. Edward O. Wilson, *Sociobiology: The New Synthesis* (Cambridge, MA: Harvard University Press, 2000).

5. I have eliminated a few layers of quotation marks here for the sake of readability.

6. I discuss this book in chapter 13, "My Science Wars."

7. Steven Pinker, *How the Mind Works* (New York: W. W. Norton, 1997), 528.

8. Joseph Carroll, "Steven Pinker's Cheesecake for the Mind," *Philosophy and Literature* 22, no. 2 (1998).

9. "Adaptationist Literary Study: An Emerging Research Program," *Style* 36 (2003): 596–617. A version of this article also appears as the introduction to Carroll's collection, *Literary Darwinism: Evolution, Human Nature, and Literature* (New York: Routledge, 2004).

10. *Evolution and Literary Theory* (Columbia: University of Missouri Press, 1995).

11. "Wilson's *Consilience* and Literary Study," *Philosophy and Literature* 23, no. 2 (1999). This is an especially brilliant and densely substantiated review of E. O. Wilson's book. More accessible is the reprint of this essay in *Literary Darwinism* (see p. 82).

12. "Human Universals and Literary Meaning," *Interdisciplinary Studies* 2 (2001). Collected in *Literary Darwinism*.

13. "Pluralism, Poststructuralism, and Evolutionary Theory," *Academic Questions* 9, no. 3 (Summer 1996).

14. Nancy Easterlin and Barbara Riebling, eds., *After Poststructuralism: Interdisciplinarity and Literary Theory* (Evanston, IL: Northwestern University Press, 1993).

15. Robert Storey, *Mimesis and the Human Animal: On the Biogenetic Foundations of Literary Representation* (Evanston, IL: Northwestern University Press, 1996).

16. Brett Cooke and Frederick Turner, eds., *Biopoetics: Evolutionary Explorations in the Arts* (Lexington, KY: Paragon House [An ICUS Book], 1999). Cooke has recently published a book-length application of adaptationist biopoetics, *Human Nature in Utopia: Zamyatin's We* (Evanston, IL: Northwestern University Press, 2002).

17. Kathryn Coe, "Art: The Replicable Unit," in Cooke and Turner, *Biopoetics*.

18. *Philosophy and Literature* 25, no. 2 (October 2001).

19. For an account of cultural biology see Steven R. Quartz and Terrence J. Sejnowski, *Liars, Lovers, and Heroes: What the New Brain Science Reveals about How We Become Who We Are* (New York: William Morrow, 2002).

20. "The Artistic Animal," *Lingua Franca* (October 2001): 28–37. Dissana-

yake's books are: *What Is Art For?* (Seattle: University of Washington Press, 1988); *Homo Aestheticus: Where Art Comes From and Why* (New York: Free Press, 1992); *Art and Intimacy: How the Arts Began* (Seattle: University of Washington Press, 2000).

21. Dissanayake, "Aesthetic Incunabula," *Philosophy and Literature* 25, no. 2 (October 2001): 335.

22. Dissanayake, *Homo Aestheticus*, 35.

23. Dissanayake, "'Making Special'—An Undescribed Human Universal and the Core of a Behavior of Art," in Cooke and Turner, *Biopoetics*, 30.

Chapter Seventeen: Ecocriticism's Big Bang

1. See "Donald E. Brown's List of Human Universals," in the appendix to Steven Pinker's *The Blank Slate: The Modern Denial of Human Nature* (New York: Viking, 2002). Also see Donald E. Brown, *Human Universals* (New York: McGraw-Hill, 1991).

2. See Pinker's *The Blank Slate*; Daniel C. Dennett, *Freedom Evolves* (New York: Viking, 2003); Stephen R. Quartz and Terrence J. Sejnowski, *Liars, Lovers, and Heroes: What the New Brain Science Reveals about How We Become Who We Are* (New York: William Morrow, 2002); William H. Calvin, *A Brain for All Seasons: Human Evolution and Abrupt Climate Change* (Chicago: University of Chicago Press, 2002); Matt Ridley, *Nature via Nurture: Genes, Experience, and What Makes Us Human* (New York: HarperCollins, 2003).

3. Jerome H. Barkow, Leda Cosmides, and John Tooby, eds. *The Adapted Mind: Evolutionary Psychology and the Generation of Culture* (New York: Oxford University Press, 1992).

4. *Practical Ecocriticism: Literature, Biology, and the Environment* (Charlottesville: University of Virginia Press, 2003).

5. See chapter 13, "My Science Wars."

6. These have been collected in Joseph Carroll, *Literary Darwinism: Evolution, Human Nature, and Literature* (New York: Routledge, 2004).

7. Joseph Meeker, *The Comedy of Survival: Literary Ecology and a Play Ethic* (3rd ed. of *The Comedy of Survival: Studies in Literary Ecology,* 1974; Tucson: University of Arizona Press, 1997).

Chapter Eighteen: Overcoming the Oversoul

1. A. Lawrence Buell, *Emerson* (Cambridge, MA: The Belknap Press of Harvard University Press, 2003).

2. Beyond the titles of the essays, it doesn't serve a useful purpose to supply

page numbers for quotations from Emerson, given the large number of existing editions, no one of which is likely to be THE one in general use. My advice to a novice reader of Emerson would be to use the Library of America *Essays and Lectures.*

3. David Mikics, *The Romance of Individualism in Emerson and Nietzsche* (Athens: Ohio University Press, 2003), 1.

4. Barbara Packer, "Origin and Authority: Emerson and the Higher Criticism," in *Reconstructing American Literary History*, ed. Sacvan Bercovitch (Cambridge, MA: Harvard University Press, 1986), 71.

5. Emerson, "Fate," in *The Conduct of Life* (1860).

6. Passages from Kierkegaard are taken passim from *Concluding Unscientific Postscript, Preparation for a Christian Life,* and *Fear and Trembling.*

7. It is worth calling attention here to Hans-Georg Gadamer's *Truth and Method,* still perhaps the definitive work on the historicity of the past and the impossibility of understanding it in its own terms.

8. For a more sympathetic account of Kierkegaard, written during the years when Joyce Carol Oates and I were arguing about Emerson, see Harold Fromm, "Emerson and Kierkegaard: The Problem of Historical Christianity," *Massachusetts Review* 9, no. 4 (Autumn 1968), 741–52.

9. Rüdiger Safranski, *Martin Heidegger: Between Good and Evil* (Cambridge, MA: Harvard University Press, 1998), 166. See also Harold Fromm, "Wrestling with Heidegger," *Hudson Review* 51 (Winter 1999): 681–90.

10. John A. T. Robinson, *Honest to God* (Philadelphia: Westminster Press, 1963), 55, 74, 76.

11. Thomas J. J. Altizer and William Hamilton, *Radical Theology and the Death of God* (Indianapolis: Bobbs-Merrill, 1966), 40, 47.

12. *Virginia Quarterly Review* 76, no. 2 (Spring 2000): 299–312.

13. Robert Denoon Cumming, *Starting Point* (Chicago: University of Chicago Press, 1979).

14. Kenneth S. Sacks, *Understanding Emerson: "The American Scholar" and His Struggle for Self-Reliance* (Princeton, NJ: Princeton University Press, 2003).

15. Laura Dassow Walls, *Emerson's Life In Science: The Culture of Truth* (Ithaca, NY: Cornell University Press, 2003), 63, 62.

16. Moncure Daniel Conway, *Autobiography, Memories and Experience* (Boston: Houghton, Mifflin, 1904), 1:281.

17. Moncure Daniel Conway, *Emerson at Home and Abroad* (Boston: J. R. Osgood, 1882), 154, 157.

18. *The Letters of Ralph Waldo Emerson,* ed. Ralph L. Rusk (New York: Columbia University Press, 1939–45), 5:195.

19. Joseph Warren Beach, "Emerson and Evolution," *University of Toronto Quarterly* 3 (1934): 474–97.

20. This is quoted by Beach (on page 482) from a lecture that Emerson gave in 1833, "The Relation of Man to the Globe."

21. *The Later Lectures of Ralph Waldo Emerson,* ed. Ronald A. Bosco and Joel Myerson (Athens: University of Georgia Press, 2001), 2:97–98.

22. Laura Dassow Walls, " 'If Body Can Sing': Emerson and Victorian Science," *Emerson Bicentennial Essays,* ed. Ronald A. Bosco and Joel Myerson (Boston: Massachusetts Historical Society/Charlottesville: University of Virginia Press, 2006), 361, 359, 360.

23. For the background to this, see the first part of chapter 16, "The New Darwinism in the Humanities."

24. James, like Nietzsche, was a more-than-casual admirer of Emerson. In 1903 he delivered a eulogizing address on the hundredth anniversary of Emerson's birth.

25. "Unhandselled" seems here to mean "untested," "untried," "primitive."

26. Murderous Viking warriors.

27. Since writing this chapter, I have been in frequent contact with Joyce Carol Oates, who reports, "I have long ago come around to admiring Emerson, very much! In fact about ten years ago I wrote a play titled *The Passion of Henry David Thoreau* in which Emerson of course appears." So my surmise seems to have been correct.

Chapter Nineteen: Back to Bacteria

1. Richard Dawkins, *The Ancestor's Tale: A Pilgrimage to the Dawn of Evolution* (Boston: Houghton Mifflin, 2004).

2. Richard Dawkins, *The Blind Watchmaker* (1986; repr. with a new introduction, New York: W. W. Norton, 1996), 112.

Chapter Twenty: Muses, Spooks, Neurons, and the Rhetoric of "Freedom"

1. See Daniel C. Dennett's *Consciousness Explained* (Boston: Little, Brown, 1991) for an extensive treatment of this subject. Dennett's more recent *Freedom Evolves* (New York: Viking, 2003) is an unsuccessful rehash of his earlier writings, described by critics as "bait and switch" because of his quirky redefinition of free will to enable him to fudge the whole issue, on which we are in substantial agreement, minus the fudge.

2. Daniel M. Wegner, *The Illusion of Conscious Will* (Cambridge, MA: MIT Press, 2002).

3. David M. Eagleman, "The Where and When of Intention," *Science* 303 (February 20, 2004): 1144–46.

4. The account that follows is heavily indebted to Gerald M. Edelman, *Wider Than the Sky: The Phenomenal Gift of Consciousness* (New Haven, CT: Yale University Press, 2004). But the enduring influence of several other books has unwittingly commandeered my neurons: Dennett, *Consciousness Explained*; Steven Pinker, *How the Mind Works* (New York: Norton, 1997); and Edward O. Wilson, *Consilience: The Unity of Knowledge* (New York: Knopf, 1998).

5. *Wider Than the Sky*, 44.

6. *Consciousness Explained*, 134–35.

7. Thomas Henry Huxley, "On the Hypothesis That Animals are Automata, and Its History," in *Collected Essays* (London: Macmillan, 1893–94; New York: Greenwood Press, 1968), 1:245.

8. "The Psychological Foundations of Culture," in *The Adapted Mind: Evolutionary Psychology and the Generation of Culture,* ed. Jerome H. Barkow, Leda Cosmides, and John Tooby (New York: Oxford University Press, 1992).

9. *Hudson Review* 43 (Summer 1990): 245–56.

10. *Hudson Review* 51 (Spring 1998): 65–78.

11. The subject of robots is taken up in the conclusion of this book: "My Life as a Robot."

Chapter Twenty-One: John Searle and His Ghosts

1. See chapter 22 for an account of Dennett and his major books.

2. John Searle, *Mind: A Brief Introduction* (New York: Oxford University Press, 2004).

3. Geoffrey Miller, *The Mating Mind: How Sexual Choice Shaped the Evolution of Human Nature* (New York: Anchor Books, 2001).

4. David Sloan Wilson, *Darwin's Cathedral: Evolution, Religion, and the Nature of Society* (Chicago and London: University of Chicago Press, 2002).

Chapter Twenty-Two: Daniel Dennett and the Brick Wall of Consciousness

1. Daniel C. Dennett, *Sweet Dreams: Philosophical Obstacles to a Science of Consciousness* (Cambridge, MA: MIT Press, 2005). (Henceforth, *SD*).

2. Daniel C. Dennett, *Consciousness Explained* (Boston: Little, Brown and Company, 2001), 454. (Henceforth, *CE*.)

3. *SD*, 12.

4. Gerald M. Edelman and Giulio Tononi, *A Universe of Consciousness* (New York: Basic Books, 2000), 158, 208.

5. Much more helpful than Chalmers's *The Conscious Mind* (New York: Oxford University Press, 1996) is his brief and pointed article, "Facing Up to the Problem of Consciousness," published in 1995 in the *Journal of Consciousness Studies* 2, no. 3:200–219. It can also be viewed online at http://consc.net/papers/facing.html. Most of the passages quoted from Chalmers are from the article as it appears in the journal, identified as *JCS* with page number.

6. See chapter 20 of this book, "Muses, Spooks, Neurons, and the Rhetoric of 'Freedom.'" See also the conclusion, "My Life as a Robot."

Chapter Twenty-Three: The Crumbling Mortar of Social Construction

1. *PMLA*, January 2006: 297–98.

2. See chapter 4, "Air and Being: The Psychedelics of Pollution."

3. This issue of *New Scientist* is variously identified as January 12 or January 13, 2007, but that the article appears on page 15 of issue 2586 seems unequivocal.

4. After this chapter was written, Mark Johnson published *The Meaning of the Body* (Chicago: University of Chicago Press, 2007). This book, a more concentrated development of themes in earlier books by Johnson alone and with his coauthor George Lakoff, offers a deeper look at the radically physical basis of neuroscience and psychology.

naturalism: biologic, 172, 178, 188, 217, 249–50; dualism, 114, 249, 260; eco-criticism, 112, 196–97

natural selection, 158, 169, 172–73, 178, 197, 217, 229, 251, 271

Nature: Darwinism, 177–88; esthetics, 8, 42, 71, 82, 119–21; human connection, 38, 42, 65, 70, 85, 95, 116, 190, 196, 217; intrinsic interest/value, 42, 67–74, 79–83, 101–2, 111–14, 122, 128–29, 180, 196; Paglia's, 152–55; technology cohabitation, 63, 112, 121–22; transcendence vs. obsolescence, 20, 38–47, 59, 61

nature vs. nurture, 164, 219–20

Nelkin, Dorothy, 143–44

neurons: consciousness, 234–37, 249, 254, 266, 274; data storage, 233–36, 260; freedom, 243–44, 261, 273, 276; writing, 237–42, 255

neuroscience, 3–5; consciousness, 191, 229, 234–37, 247–49, 254–55, 272, 274; Darwinism, 159, 169–76; eco-criticism, 190–93

New Age, 66, 69, 74, 98, 140

New York City, 14–15, 19, 22, 33, 51, 110, 271

Nietzsche, Friedrich, 151, 155, 167–68, 202, 207–8, 217

non-life on earth, 229, 256, 258

"normality," human, 271–72, 277

North Barrington (Illinois), 19, 118

Oates, Joyce Carol, 198, 201, 212, 219–20

obsolescence, transcendence vs., 20, 38–47, 59, 61, 277

Oelschlaeger, Max, 110, 119

Olympics, air pollution, 1–2

omnipotence, human, 44, 65

organisms, primal, 225, 228

Other, speaking for, 101–2, 111, 128–29, 132

Oversoul, Emerson's, 198–220

ozone level, 1, 17, 43, 49–52, 77

Paglia, Camille, 87–88, 146–56

particulates, airborne, 17, 27, 51

pastoralism. *See* farmland / farm life

performance, physical, pollution impact, 17–19, 25–26, 32–33, 37, 52–53

personalities. *See* human attributes

personal realities: constructionism, 264–69; freedom, 242–46, 261–62; pollution, 25, 28–32, 45, 49–55, 98

pesticides, 42, 55, 118

phenomena, real world, 162; consciousness, 72, 195, 249–50; knowledge construction, 139, 179–81, 189, 244; supernaturalism vs., 114, 161, 209, 211, 229

philosophy: consciousness, 9, 159, 229, 234–35, 248–49; existential, 198–220; free will, 219, 248–52; linguistics, 180, 188, 248; self, 128, 159, 248, 251–52

Phoenix (Arizona), 105, 110, 119–20

physical life. *See* life on earth

physical symptoms, pollution, 1–3, 13–19, 35, 54; data sources, 49–50, 54–55; testimony, 25–34, 45, 52–57, 97–98, 105–6, 110, 271

physical systems: experience source, 260, 262, 274, 276; mental states, 249–50, 266

physico-chemical-biological life, 166, 229, 276

Pinker, Steven, 6, 167, 170–77, 195, 219, 247, 263

plants, retro-history, 225, 228

Plath, Sylvia, 238–40

Plato, 128, 167–68, 205

play: free, intellect and, 140, 169, 181, 242–43; realism, 125

pleasure seeking, 178

PMLA, 135–37

political correctness: modern, 168–71, 184; Paglian, 150–53; postmodern, 87, 112; realism, 125–27, 131–32; science, 144–45

politics: ecology, nature/social, 66–68, 71–76, 83, 277; environment reform, 1, 27–31, 96, 99–100, 122; feminism, 135–36, 182; foreign, 225–26; sexual, 268–69

pollution, 1–2, 9–10; awareness vs. un-awareness, 10–35, 45, 48–57; rural

pollution *(cont.)*
  vs. urban, 13–19, 22–24, 32–34, 107,
    118, 122; watches and warnings, 17,
    35, 49. *See also* air pollution; water
    pollution
pollution drift, 17–18, 22–26, 29, 37,
    49–51
pop stars/culture, 148, 153–54, 235, 240
postmodernism: ecology, 85–94, 112,
    116, 135–37, 146, 155, 161; fiction,
    91, 123–32
preferences, human, 160, 170, 173, 176
preservation, 66–67, 70–71, 82
professionalism, 165–66
psychological symptoms, pollution, 17,
    19, 32–34; testimony, 53–54, 271
psychology: evolutionary, 170, 177,
    189–90, 193–96, 235, 244–45; exis-
    tential, 208, 276; freedom, 169, 203,
    213, 250, 261–62
public policy. *See* reform ecology

race/ethnicity: ecocriticism, 90, 189, 192,
    226–27; realism, 126–27, 131–32, 243
radical ecology, 66–68, 102, 138
rationality: alternate, 97–98; body vs.
    mind, 38–39, 234, 250, 277; deep
    ecology, 65, 75, 102–3; environmen-
    talism, 45–47, 88, 93; knowledge,
    128–29, 202; science, 138–45, 217
realism: animal literature, 123–32; eco-
    logical, 88–91, 197; existentialism,
    213–14
reason. *See* rationality
recycling, 111, 190–91, 212, 216
reflection, conscious, 111–12, 260, 275
reform ecology, 1, 30–31, 66–71, 76;
    regulations, 96, 99–100
religion, sociobiological roots of, 160–61,
    229. *See also* Christianity
Reno (Nevada), 104, 114
reproduction: behavior, 172, 182–83,
    223, 251, 268–69; genetic, 174, 193,
    251; retro-history, 222–29
resources: biocentrism, 70, 74, 78–81;
    destruction, 67, 74, 130, 196–97;
    instrumental value, 83–84, 101
respiratory illness, 1, 15, 19, 25–26, 28,
    30, 35, 45, 49, 52–53

revisionist theology, 206, 211–12
ring species, 226
ritualization, 74, 187
roadways: accidents, 18, 32–33; air pol-
    lution, 1, 14–15, 23, 54; Interstate
    trip, 1, 63, 104–17
Robbins, Bruce, 137, 145
Robinson, John, 208–10
Robo-Mary, 258
robots, human, 246, 256, 261, 271–77
romanticism, 203–6, 213, 217
Rose, Hilary, 143
Ross, Andrew, 136–37, 139–42, 144–45

Sand County Almanac, A (Leopold), 78,
    80–82
sanity/insanity, 277
San Joaquin Valley, 1–2, 104, 106–7
Scholes, Robert, 264
science: academia and, 135–45, 165–66;
    consciousness, 247–48; ecology, 43,
    65; existentialism, 214, 217; reduc-
    tionism, 161–62, 260; rhetoric, 162,
    179–80, 189; validity, 5, 30, 144, 162
science wars: academic, 136, 141, 144–
    45, 161, 165–66; cultural, 135–45;
    humanities, 4–9, 158–60, 175–79,
    195, 197, 269; political, 136, 140–41,
    145, 155
Science Wars (special issue of *Social
    Text*), 136–39
science, male, 138–39, 142
Searle, John, 247–52, 262
self: contemporary, 159, 245–48, 251–52;
    Darwinism, 168, 173–74; ecocriticism,
    91, 190–93; Emerson's, 119–220; ex-
    pressions, 3, 7–9, 164–68, 261, 277;
    imaginary, 234, 255; Imperial, 95–96,
    238, 240; Other vs., 101–2, 111, 128–
    29, 132; virtual, 234–35, 238–42,
    248, 251–52, 255, 266
selfishness, 80, 165–66, 174, 197
self-reliance, 198, 201, 205, 208, 212
"sense of being": drugged from pollu-
    tion, 25, 30–35, 52–55, 107; in real-
    ism, 126–28, 132
senses: constructionism, 175–76,
    243–44; pollution, 23–33, 51–52, 55
Sessions, George, 79

Credits

Earlier versions of the following chapters in this book appeared in the publications shown below.

Chapter 2. *The Yale Review* 65 (1976): 614–29.

Chapter 3. *Georgia Review* 32 (1978): 543–52.

Chapter 4. *Massachusetts Review* 24 (Autumn 1983): 660–68.

Chapter 6. *Hudson Review* 45 (Spring 1992): 23–36.

Chapter 7. *ISLE* 1.1 (Spring 1993): 43–49.

Chapter 8. *Hudson Review* 48 (Winter 1996): 691–99.

Chapter 9. *Electronic Book Review* 8 (Winter 98/99), www.altx.com/ebr/reviews/rev8/r8fromm.htm.

Chapter 10. *Hudson Review* 51 (Spring 1998): 65–78.

Chapter 11. *Holding Common Ground: The Individual and Public Lands in the American West,* ed. Paul Lindholdt and Derrick Knowles (Spokane: Eastern Washington University Press, 2005), 36–40.

Chapter 12. *Hudson Review* 53 (Summer 2000): 336–44.

Chapter 13. *Hudson Review* 49 (Winter 1997): 599–609.

Chapter 14. *Hudson Review* 48 (Summer 1995): 308–16.

Chapter 15. "A Crucifix for Dracula: Wendell Berry Meets Edward O. Wilson," *Hudson Review* 53 (Winter 2001): 657–64.

Chapter 16. Part 1, "From Plato to Pinker": *Hudson Review* 56 (Spring 2003): 89–99. Part 2, "Back to Nature, Again": *Hudson Review* 56 (Summer 2003): 315–27.

Chapter 17. *Logos* (Summer 2004), http://logosjournal.com/fromm.htm.

Chapter 18. *Hudson Review* 57 (Spring 2004): 71–95.

Chapter 19. *Hudson Review* 58 (Autumn 2005): 519–27.

Chapter 20. *New Literary History* 36 (Spring 2005): 147–59.

Chapter 21. A review of *Mind: A Brief Introduction,* by John Searle (New York: Oxford University Press, 2004), in *Georgia Review* 59 (Fall 2005): 716–20.

Chapter 22. *Hudson Review* 59 (Spring 2006): 161–68.